New Zealand

New Zealand

A NATURAL HISTORY

Tui De Roy & Mark Jones

FIREFLY BOOKS

A FIREFLY BOOK

Published by Firefly Books Ltd. 2006

First printing

Publisher Cataloging-in-Publication Data (U.S.)

De Roy, Tui.
New Zealand : a natural history / Tui de Roy and Mark Jones.
[160] p. : col. photos. ; cm.
Includes index.
Summary: Illustrated survey of the geography, flora and fauna of New Zealand.
ISBN-13: 978-1-55407-196-8
ISBN-10: 1-55407-196-8
1. New Zealand—Geography. 2. Plants—New Zealand. 3. Animals—New Zealand. I. Jones, Mark. II. Title.
508.93 dc22 QH197.5D37 2006

Library and Archives Canada Cataloguing in Publication

De Roy, Tui
New Zealand : a natural history / Tui de Roy and Mark Jones. Includes index.
ISBN-13: 978-1-55407-196-8
ISBN-10: 1-55407-196-8
1. Plants—New Zealand. 2. Animals—New Zealand. 3. New Zealand—Geography. I. Jones, Mark II. Title.
QH197.5.D47 2006 508.93 C2006-900516-8

Published in the United States by
Firefly Books (U.S.) Inc.
P.O. Box 1338, Ellicott Station
Buffalo, New York 14205

Published in Canada by
Firefly Books Ltd.
66 Leek Crescent
Richmond Hill, Ontario L4B 1H1

First published in 2006 by
David Bateman Ltd.
30 Tarndale Grove, Albany
Auckland, New Zealand

Design: Think Red
Editorial: Andrea Hassall
Printed in China by Everbest Printing Co.

For Julian and Julie,
who, individually, have brought
an infinitude of new possibilities to
our personal life stories.

AUTHORS' NOTES

All photographs in this book were taken by Mark and Tui either in free and wild conditions or, in some cases, in breeding facilities operated expressly for conservation purposes (e.g., kakapo and great spotted kiwi). No images have been retouched digitally, double exposed or otherwise altered. Fuji films and Nikon equipment was used throughout. Although the writing is co-authored, we have divided the primary responsibilities, and hence the first person tense, as follows: Tui penned the main chapters and conceived the layout; Mark produced the conservation introduction and the index.

The names used for New Zealand species are those in common usage and, where these are English, the Maori name (if one exists) is also stated the first time the animal or plant is mentioned. Maori names are not pluralized as in the English language (e.g., one kiwi, two kiwi). A full list of English, Maori and Latin names can be found in the appendix.

PAGE 1 North Island brown kiwi running on the forest floor.

PAGE 2–3 Winter sunset tinges heavy ash cloud over an erupting Mount Ruapehu.

ABOVE Kiwi chick feather detail.

OPPOSITE Okarito kiwi, a new species known as rowi, scampers beneath ferns and rimu seedlings in Westland National Park.

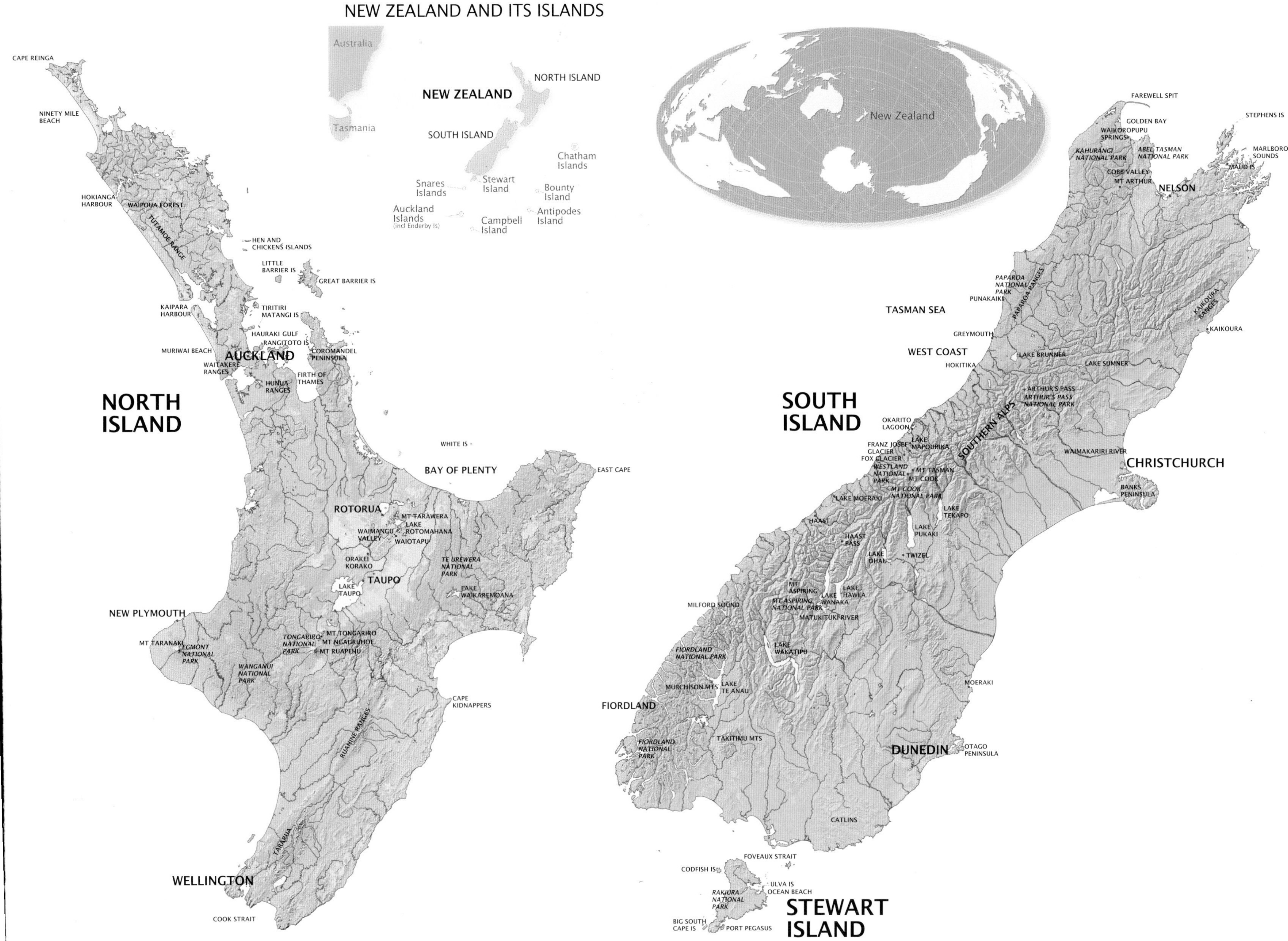

LEFT Kea interact playfully amid dwarf alpine flowers above Fox Glacier in the Southern Alps.

Acknowledgements

A group of very large islands with only four million people, in many ways New Zealand feels like the metaphorical village, where like-minded people communicate and reach out, become enthused with one another's projects and offer assistance and knowledge freely. Nowhere is this trait more strongly felt than in conservation circles. United by a vision of wild and pristine nature, and the urgency to halt and reverse the damage that humans have wrought upon these shores in very recent history, the network of New Zealand conservationists spreads something like a family across the furthest points of this young nation. So it may not come as a surprise that a large number of people have been directly or indirectly involved in helping us create this book; sharing wisdom, knowledge, guidance and companionship in the field, all with equal generosity.

For our somewhat dogged interest in the secretive ways of the kiwi, we owe our motivation to fellow photographer Rod Morris, who offered valuable yet unprompted suggestions. Later, a bevy of kiwi workers in the field entrusted us with some of their most precious charges. We thank Chris Rickard — and his kiwi dog Gem — Selena Brown, Joe Crofton, Rob McCallum and Jo Macpherson in Franz Josef, Phil Tish and Crissy Wickes in Haast, John Lyall in Hokitika, Hugh Robertson and Rogan Colburn in Wellington, to name but a few. The Otorohanga Kiwi House opened its doors to us most generously, allowing us intimate encounters with their rare breeding birds, for which we thank our old friend Warwick Reed, along with Eric Fox and Dominique Fortis.

Photographic colleagues — some whom we have yet to meet — have rallied with information and advice. For their spontaneous sharing of knowledge we thank Kim Westerskov, Darryl Torckler, Andy Belcher, Gideon Climo, Kennedy Warne and Warren Judd, editor of the superb bimonthly *New Zealand Geographic* magazine.

Kakapo feather detail.

Longtime friends Colin and Betty Monteath provided insights, as well as kindling our interest in the mountains. Roger Sutherland and Barbara Todd sharpened our enthusiasm for the wildlife off our shores. Peter Fullerton shared our memorable first sailing adventures in New Zealand waters. Neville Peat helped focus our minds on the crafting of words, whereas Andris Apse, Rod Morris, Geoff Moon, Rob Brown and Rob Suisted all have inspired our images. Bruce Watson opened doors and gave us the confidence to proceed. Tracey Borgfeldt, our publisher, invested time and effort far beyond any call of duty.

Staunch supporters of our work throughout is the hardy band of Department of Conservation (DoC) officers scattered across the nation but united in the energy they spend on protecting beleaguered species. The Southern Islands team, led by Greg Lind, Pete McClelland, Jeremy Carroll and later Andy Roberts, has been steadfast. In the north, the Supporters of Tiritiri Matangi Island, and in particular DoC rangers Ray and Barbara Walters, were warm and invaluable. Richard Maloney provided help with black stilts in the Mackenzie Basin. Paul Jansen was most supportive and his entire Kakapo Recovery Team on Codfish Island truly welcoming. Locally, Greg Knapp and Simon Walls pointed us to the falcon's eerie, while Peter Gaze opened our eyes to the world of frogs, geckoes and weta on Maud Island, where the resident rangers Bron and Damon shared the magic of an unravaged microcosm.

Like us, many of our friends do not distinguish between personal and professional insights. Thank you Colin and Betty, Pete, Greg and Sue, Ali, Bruce and Claudia, Paul, Mike and Carol, Jacinda, Grant, Carl, Chris and Sarah; for being there.

OPPOSITE Lush ferns adorn the forest floor, Ulva Island Sanctuary.

CONTENTS

OPPOSITE A week-old Okarito kiwi, rowi, chick ventures out alone to explore its secretive world.

Preface

Takahe feather detail.

By its name, New Zealand sounds like something novel. From the human perspective, this young island nation came along very recently indeed. Discovered and settled within the last millennium or so, first by the original Maori people of Polynesian ancestry, who called the land Aotearoa — 'land of the long white cloud' — and some centuries afterwards by Europeans, her wild spirit has barely been tamed. Yet from a natural perspective this primordial chunk of earth crust found its tectonic and biological identity long before the other land masses as we know them had even been born.

Eighty million years ago dinosaurs were breathing their last and proto-mammals had barely started cutting their teeth in the Cretaceous forests of the day. At the same time a profusion of birds — direct cousins of those expiring dinosaurs — thrived, sharing space with shy, primitive reptiles and frogs. About then, as the ancient southern supercontinent of Gondwana was fragmenting into separate land masses, one sliver broke away and drifted out alone into the vastness of the open ocean. Riding on this geologic-scale Noah's Ark, a sample of the primal ecosystem thus departed on an independent journey that continues today. A few bats were the only mammalian representatives that belatedly caught up to join a living community where birds would evolve to rule.

In time, and in the absence of other land mammals, these birds would take on the roles of grazers and browsers, predators and pollinators. And so it was that a living time capsule now called New Zealand entered the Eocene age — the current era — with a bevy of organisms that almost defy imagination. Giant eagles soared on 3-m (10-ft) wingspans, frogs carried their young on their backs, carnivorous snails and multi-legged worms prowled the forest floor. And a whole range of land birds — in fact more than half of all New Zealand birds — lost their ability to fly in favour of a ground-dwelling existence. Flightless parrots, flightless geese, flightless gallinules and even flightless wrens scampered from seashore to mountaintop. Eleven species of moa — distant relative of ostriches, as is the kiwi — divided the habitat into a preposterous world of feathered pedestrian giants. There were stout-legged moa, heavy-footed moa, upland moa, slender bush moa, large bush moa, crested moa and the giant moa standing over 3 m (10 ft) tall, as well as a bizarre stocky lizard hunter, named the adzebill, that was related to no-one in particular.

When humans finally arrived, they did not prove to be very careful or considerate custodians of this unusual microcosm. Within a very short time, fire and hunting spelled the extinction of 25 out of 40 flightless New Zealand birds. With the later arrival of European settlers, a second wave of devastation occured, as introduced mammals invaded the landscape and forests were razed for agriculture. This time, even small flying birds were not spared.

But the extraordinary uniqueness of the New Zealand/Aotearoa fauna endured even after these losses. Today, from one end of the country to the other, teams of tenaciously dedicated men and women are battling against rats, stoats, possums and goats, European gorse, South American passion vine and Asian honeysuckle, to name but a few of the invasive species that are threatening the continued existence of New Zealand's own weird and wonderful species.

In these pages we would like to sing the praises of their successes, warn against the risks that lie ahead, and above all celebrate the natural treasure that New Zealand holds. To do so we share our explorations of this natural world from an intimate perspective. It is through the complex lives of the many rare species, large and small, that we would like to impart our wonder for this extraordinary wild realm. This is a journey where the eons past echo each dawn in the mesmerising chorus of bellbird and tui, kokako and kaka; are reflected in the crystal sunrises over immaculate snowy mountains; felt in the rumble of erupting volcanoes and shuddering earthquakes; and smelled in the musty aroma of misty, moss-spangled podocarp forests.

Tui De Roy and Mark Jones

OPPOSITE Mist hangs in the splendid beech forest of the Haast Valley, Westland National Park.

INTRODUCTION
PRESERVING THE KIWI'S FOOTSTEPS
Essay by Mark Jones

I feel as though I could be watching the very dawn of time; the way it has always been. Here, in winter especially, it's often the way it still is, but only for those who make the effort, forsake a bit of sleep and doggedly shun the disappointment of frequent failure. But then, how could watching the awakening day ever be disappointing? And I know that this sunrise is going to be very special.

The sky and rippled clouds are being suddenly painted with vibrant reds, burnt oranges and glowing pinks, yet the span of the eastern horizon remains perfectly clear, with a narrow band of paling velvet blue representing the infinitudes of space and the beyond. I feel sure that in just a few short minutes I will witness one of the most ephemeral and, for the sceptics, controversial atmospheric phenomena. Entire scientific papers have been written about it, defining its technical parameters, trying to convince the unconvinced in the custom of pure science. Yet this will be a practical exposé, I feel sure.

Minutes seem to linger for an eternity, as is the way for those expecting miracles, but if it happens now it will be for the tenth day in succession. Then, just as I am tempted to glance away at some minor distraction — an oft-repeated mistake at this point — the first blazing rays of the rising sun make their debut on the new day and, briefly, there is no doubting the spectacle. As the very first rim of that life-giving star emerges out of the ocean, it appears as a brightly glowing green orb before quickly assuming its normal intense golden hue, ascending gradually into the sky and distortedly breaking itself away from the horizon to free the land from the chilled confines of the long night.

ABOVE The kiwi's footsteps, still prevalent on Stewart Island's Ocean Beach, are becoming scarcer due to introduced predators.

OPPOSITE Sun sinking into the sea, West Coast. The extreme clarity of the New Zealand air produces memorable sunrises and sunsets.

The so-called Green Flash — though not illusory, perhaps in fairness "green blip" would be a better term — is an optical phenomenon created under certain atmospheric conditions by which the sun's rays are for an instant refracted in rainbow-fashion above the horizon. Occasionally the effect can even manifest deep blue or mauve instead of green. It is more often seen at sunset, just as the last arc of the sun's rim disappears, but that is because more people tend to watch for it then. Those who have witnessed it can almost be likened to religious converts in their desire to convince those less-fortunate, who have yet to 'see the light'. In the case of the Green Flash, seeing is believing, and such an astronomical demonstration can certainly make you start the day with a smile.

But one does not have to look to the universe to behold the unusual. There is so much more to be seen and believed right here in the frequently bizarre nature of New Zealand.

The exceptional clarity of the air is but one example of the many features of these islands that attract superlatives. Whether one talks of reptiles, birds, amphibians, insects or plants, terms like largest, smallest, tallest, the world's only, southernmost, most ancient, longest-lived, unique, rarest, heaviest, oldest, fastest growing… the litany seems to know no limits: a comment like 'most unusual' becomes almost commonplace. Flightlessness and gigantism abound, but so too does dwarfism. Being nocturnal is almost *de rigueur.* There are many cases of high diversity of similar but distinct species, yet nearly everything alive represents rarity, from a bird that mates face to face (stitchbird) to frogs with webless feet and no tadpoles (Archey's), from giant insects that behave like rodents (weta) to a predatory snail the size of an English muffin. Set amongst these creatures a nocturnal flightless bird with long whiskers and fur-like feathers that finds its food by sense of smell (the kiwi) really doesn't seem so offbeat.

OPPOSITE Pohutukawa in full summer bloom bedeck the shores of Tiritiri Matangi Island, Hauraki Gulf.

The weathered stump of a massive totara tree, overlooks Golden Bay at the top of the South Island; a forlorn reminder of ancient forests turned to sheep pastures.

Though endearing in their native land, Australian brush-tailed possums are laying waste to New Zealand forests, consuming an estimated 70,000 tons of vegetation each night.

The first of the two most significant events in the history and evolution of species on the islands of New Zealand was that indefinable moment some 80 million years ago when continental drift effectively rendered them inaccessible to all but marine or flighted oceanic organisms. In effect, New Zealand was cut off from the rest of earth's evolutionary pathways. Thus, spared from the emergence of terrestrial mammals, these islands became a Gondwanian relict, a land that has been described as both a Noah-like ark and our best chance of studying an experiment in independent evolution without resorting to another planet.

In this land that has weathered multiple and severe glaciations, drastic rises and falls of sea level, upheavals of mountain chains and the violent birth and rebirth of volcanoes, the animals and plants have stolidly endured the evolutionary bottlenecks forced upon them, successive generations adapting and flourishing in benevolent innocence. The statistics for endemic species (those found nowhere else) are astounding; 100 percent of reptiles, 100 percent of amphibians, 90 percent of insects, 87 percent of terrestrial birds, 83 percent of freshwater fish, 81 percent of plants, even 44 percent of breeding seabirds. Globally, this proportion of endemism is second only to the islands of Hawaii.

The second most influential, and ultimately devastating, event happened barely a millennium or two ago when humans finally ventured across the isolating oceans in search of a new Utopia.

Undoubtedly the first mammalian feet to step onto these shores were human, when Polynesian seafarers, the ancestors of Maori, leapt through the surf from their waka (ocean-going canoes). Within minutes kuri (their hunting dogs) would have followed them, to be joined shortly thereafter by the kiore (Pacific rats) they carried as a reliable food source. And so these rootless islands lost their innocence, their insular ecosystems belatedly forced to join the model set for the rest of the mammal-burdened planet.

It appears that the very first people did not stay long, or maybe they just didn't survive, but there is evidence that some 2000 years ago someone did leave rats to make their first inroads into the bird-dominated food chain. New waves of waka-borne Polynesian settlers some 800 to 1200 years ago (to which all modern Maori can justly trace their lineage) hunted and fished their way across the length and breadth of their new homeland. Their permanence brought fire and agriculture to the landscape, so beginning a transformation — and subsequent annihilation — of habitat and species that was then accelerated by the arrival of technologically advanced Europeans in the late 1700s and 1800s. Bringing livestock and axes, and armed with big ideas and Victorian ideals, these pioneers' successive introductions of alien animals began an unprecedented onslaught. More species have been intentionally and successfully introduced to New Zealand than to any other country; spawned partly by the perception that the country exhibited a paucity of fauna. The indigenous animals — and plants — were maladapted and

OPPOSITE Three Mile Lagoon forms a classic wetland where the Cockabulla Creek spills into the wild Tasman Sea, South Island.

Under the pall of erupting Mount Ruapehu, motorists are reminded that cars take their toll of kiwi in Tongariro National Park.

Department of Conservation kiwi experts Selena Brown and Chris Rickard, with specially trained kiwi dog Gem, track the health of the rowi, a newly described species confined to the Okarito Forest, Westland.

gullible to the artful ways of the mammalian invaders, and ultimately the tumultuous changes wrought upon their world would rapidly take many of them down the road to extinction.

The islands of New Zealand were this planet's last significant habitable landmass to be colonised by people. To visualize this timing in some sort of perspective, if one can imagine the whole of the 80 million years since the break-up of Gondwana represented by the last 12 months, then the few minutes it has taken you to read from the beginning of this chapter is the approximate proportion of time that humans have been part of these islands' history. Needless to say, their effect has been of far more dramatic proportions. Of vertebrates alone, some 64 species no longer patrol their ancestral domain. More than half of them were extinguished, or verged on being so, before the main European influx. Originally covering some 85 percent of the land surface, barely a fifth of the native forests still exist. Sadly, nowhere in this archipelago can be alleged to be truly pristine, in the sense that it was before humanity arrived.

However, in the blink of that imaginary 12 months a third significant, and ongoing, event has also occurred: the birth of awareness. A profound awareness in fact, of the uniqueness of New Zealand's biological heritage, an enlightened recognition of the absolute values intrinsic to the nation's cultural and spiritual — let alone ecological — well-being. In spite of an historically exploitative outlook and abysmal track-record, an appreciation for their natural surroundings is something inbred in all New Zealanders, regardless of ancestry. These emotions run deep in the island blood, beginning with the early Maori whose tohunga — guiding elders — tried to slow the exploitation of some species through sacred inference. These

early considerations were given a darker twist by the pioneering European explorers, surveyors and farmers who made some of the more astute commentaries on the declining state of the country's flora and fauna. Through today's conservators and scientific advisors defining future management policy, attitudes are slowly changing from colonial exploitation to one of taking responsibility for the preservation of the country's natural treasures.

Everyday modern New Zealanders hold the key to biodiversity's long-term survival, an awareness born from within the culture. We have thus bestowed upon ourselves an enormous responsibility towards the conservation and preservation of our natural environment. We represent barely 0.06 percent of the world's population, yet are responsible for at least 5.3 percent of the world's endangered birds. Of New Zealand's 12 completely endemic taxonomic Orders or Families, no less than nine are officially listed as one hundred percent endangered or threatened. With well over 1000 species on the endangered list, more than 360 are considered in a nationally critical state. At least five endemic bird species have been (or still are) down to less than 50 individuals, their survival hinging on intensive and inventive management schemes. Numerous species now subsist only as refugees on specially protected islands. These statistics rank New Zealand in the world's top five countries for threatened species.

Extinct species are beyond redemption: mourning their loss, either emotionally or academically is a purely philosophical exercise. Instead, the regime is now firmly focused on learning from past mistakes and defending the abundance and diversity of animals and plants throughout all habitats. This is something that the new wave of New Zealand conservationists excels at, even though the challenges — financial and logistical — remain huge. Predator control, habitat restoration, species rehabilitation, captive breeding, hand-rearing or cross-fostering have all been applied successfully. Sophisticated techniques and novel approaches to complex problems with critically small populations in a difficult and rugged country have been trialled and implemented. The lessons learned and technologies used in New Zealand are now sought by other nations, with personnel, know-how and equipment often being exported to help save equally threatened environments elsewhere in the world.

It is a credit to the dogged commitment of the New Zealand conservation workers, toiling under marginal conditions and minimal support over dozens of years and whole careers, that one species after another is clawing its way precariously back from the very brink of the irreversible abyss. In fact, after a tidal wave of extinction within the last century, not one more species has been lost in the last 40 years. Nonetheless, if rigorous management programmes ceased today dozens of species would soon be in a dire state.

At the head of the taskforces leading these advances is the government-funded Department of Conservation (DoC). While there are nearly constant political — and indeed almost fashionable — social pressures to ration the department's jurisdiction and curb the sometimes controversial, though arguably effective, management methods it employs, the reality is that DoC-led teams (management committees,

OPPOSITE In the last 15 years a gannet colony of about 1200 pairs became established on the sea cliffs above Muriwai Beach on Auckland's west coast, North Island.

Only days old, this dead kiwi hatchling shows the telltale bite of an introduced stoat, a ravenous predator that kills about 70 percent of all chicks before they are six months old.

Wearing a tiny radio transmitter to follow its success, a newly hatched rowi chick ventures alone from the parental burrow into a forest patch painstakingly cleared of predators to increase its chances of survival. Without such help less than 5 percent survive to adulthood.

field researchers and rangers, advisors and closely associated organizations) are racking up a growing catalogue of outstanding conservation successes. Oftentimes these are accomplished on remarkably meagre and fickle finances — in 2004 DoC's annual budget was barely $NZ295 million ($US199 million), a figure the NZ Health Department spent about every 11 days.

In that same year, following a Herculean aerial poisoning operation, DoC cautiously declared subantarctic Campbell Island rat-free, the largest rodent control programme ever undertaken anywhere in the world — an area of some 11,300 ha (28,000 acres) which had the highest density of rats *ever* recorded. Consequently, as I write these words, 55 flightless teal, the world's smallest duck, endemic to Campbell Island are being released back onto their native shores. Wiped out by Norway rats introduced 200 years ago, they are descendants of just four birds that were rediscovered on a tiny storm-beaten satellite islet and bred in captivity (on predator-free Codfish/Whenua Hou Island Reserve).

Once considered a hopeless cause, reversing the devastation wrought by rats falls only just short of a miracle. Now with a population of 180 birds and the security of its island home assured, the world's rarest duck will have its extreme critical status downgraded. Such dreams nurture our well-being, our human and humane integrity. From pygmy button daisy to striped skink, weta to weka, mudfish to bat, frog to snail, kaki to kakariki, Species Recovery Plans for a whole range of beleaguered endemics — over 50 detailed conservation strategies — aim to redress

OPPOSITE Ever playful, dusky dolphins leap up in unison off the coast of Kaikoura, South Island.

DoC's Tamsin Bliss and Phil Tisch carefully check the health and weight of a rare female Haast tokoeka, possibly a new type of kiwi. Its legs are the strongest part of its body for handling.

Jason Taylor is detailed by DoC to ensure breeding success of re-introduced stitchbirds on Tiritiri Matangi Island, where parasitic mites are controlled in specially designed nest boxes and chicks are colour-banded to track their later success.

the steady desecration of New Zealand's mountains, forests, rivers, wetlands and shorelines, and maintain the biodiversity these habitats deserve. High-profile species — the tuatara, takahe, kakapo, kokako and kiwi — have each become charismatic national icons for the successful 'salvation' of species, some against near-impossible odds.

When we direct our efforts towards these most endangered and symbolic creatures, it often safeguards other aspects of their habitat and benefits other less alluring but equally valuable life forms. The Royal Forest and Bird Protection Society, through its network of regional clubs, fosters this holistic approach and helps communities deal with issues both local and national. Sponsorships by leading companies help channel much-needed extra funds, as well as keep themselves and their icon in the public eye. For the last 14 years, every time I write a cheque I am contributing a small amount via the Bank of New Zealand for the privilege of having the Kiwi Recovery emblem on my payment. Each cheque represents a minimal contribution, yet with enough people over time it adds a significant amount to the overall coffers used to safeguard the nation's avian idol.

But while there are many species that continue to struggle for this scale of recognition, a whole bevy of other businesses, from those in the travel industry to law firms, freighting companies to wineries, are coming aboard to endorse a number of programmes through the endowment initiatives of the recently formed National Parks and Conservation Foundation. For instance, through the Threatened Species Trust efforts are in place to attract sponsorship for the little stitchbird (hihi), a charming unpretentious honeyeater once found throughout the North Island forests, but whose strident call has not been heard there since 1883. Under

the campaign banner 'A Stitch In Time' — a would-be natural insignia for any clothing manufacturer or related industry — it has been calculated that as little as US$360 will save a breeding pair of this, one of the world's rarest birds. Vulnerable to cats and rats, and susceptible to introduced avian diseases, there is but one self-sustaining island population left, on Little Barrier Island Reserve. Its recently reviewed status has been upped to nationally endangered. The modest goal of DoC's recovery programme is to re-establish five self-sustaining populations.

If Comalco, the country's huge aluminium smelter, can bravely channel thousands of dollars towards kakapo breeding efforts, Mainland Cheese likewise for the yellow-eyed penguin, the Bank of New Zealand in favour of kiwi, surely a trendy fashion label should be able to contribute to the survival of the petite and dapper stitchbird — a 'stitch in time' indeed.

New Zealand is fortunate to have a generous collection of offshore islands and islets that can be run as isolated reserves with relative ease, fortified bastions to which the survivalists can safely be evacuated. Over the last 12 years some 132 reintroduction projects involving 63 species have helped mitigate the vulnerability of localised remnant populations. But in the wake of successes like the Campbell rat programme, with bold and innovative thinking and sound research, even the concept of an 'island' is being greatly expanded. Six significant areas, termed Mainland Islands, are being effectively managed to the collective benefit of the entire ecosystem they encompass. Often with the participation of local community groups and societies, these high-maintenance projects involve everything from servicing hundreds of kilometres of predator trap-lines and fencing, to ripping out invasive weeds. Threats seem ever-looming and setbacks spark resource-sapping damage control operations: rat and stoat infestations, rogue dogs, virus outbreaks and the like. Yet, after only a few years, systematic monitoring has shown a dramatic improvement in the overall health of these coveted habitats: no kiwis killed by predators; juvenile robins up from 18 percent to 53 percent; 95 percent fledging success for blue duck; first sightings of tusked weta; rare frogs rediscovered; and even rarer plants safeguarded.

OPPOSITE Under the cover of night, a kakapo scrambles through scrub in search of berries on Codfish Island.

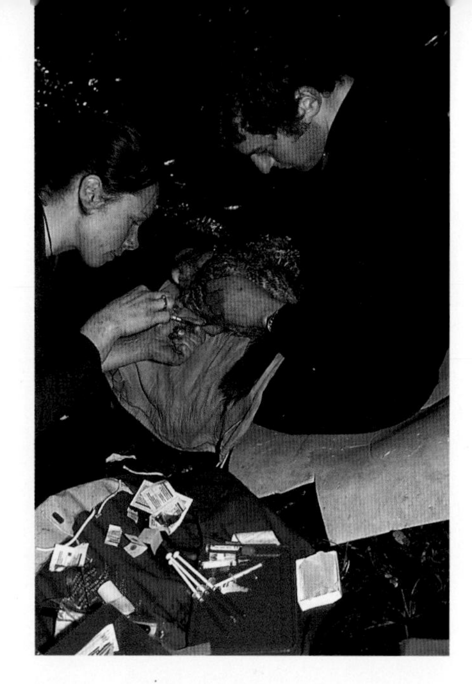

As meticulous as field surgeons, Kakapo Recovery team members Malcolm Rutherford and Dierdre Vercoe on Codfish Island perform an annual health check, including immunisation and blood tests, on the patriarch breeding male kakapo 'Richard Henry' (named after a pioneer conservationist of the 19th century), the last bird rescued from Fiordland National Park 30 years ago.

Taking up to three days to break out of the egg after more than two months incubation, a newly hatched North Island brown kiwi is gingerly lifted from the incubator at the Otorohanga Kiwi House, symbolising hope to reverse the species' decline.

In addition to these large-scale operations, through state trusts and binding covenants, the growing conservation ethic is prompting farmers and landowners to set aside considerable tracts of valuable private land, adding some 97,000 ha (240,000 acres) to the country's conservation estate. Today the wholesale clearance of native forests has ceased and more than 30 percent of the country — some 8 million ha (20 million acres) — are afforded some degree of protection. The South Island is especially well endowed, having a contiguous tract of interlinked national parks and reserves running the entire length of its mountains and western forests.

The essence of New Zealand's 80 million-year-old history still shines vibrantly before our eyes and, amazingly, new species are being discovered almost literally as you read this book. While it seems difficult not to lament what is no longer with us, only in so doing are the riches that remain brought sharply into perspective. Indeed, this insular world remained untouched by humans longer than any comparable biota on this planet, and while there is still so much that might yet be lost, this represents in equal measure what it is possible to preserve. Hopefully, our modern society will measure up as worthy custodians of such a treasured heritage.

1
THE SUNNY FOREST MARGINS
SPRING BLOOMS AND HONEYEATERS

The first hint of dawn barely tints the summer sky when I slip out of the bunkhouse on Tiritiri Matangi Island. Across the calm Hauraki Gulf, the glittering lights of Auckland remind me that the hubbub of civilisation is a mere 24 km (15 miles) distant. But as I head down to the beach, I am at once enveloped in the most wondrous, soul-lifting wild bird chorus I have ever heard anywhere on this planet. As if directed by an unseen conductor presiding over an island-wide orchestra, fuzzy-tongued nectar-lovers, ancient wattlebirds and discreet insect-eaters together greet the new day with unmitigated exuberance. One hundred, two hundred birds perhaps, each following its own, unbroken rhythm; waves of music rising and falling, pulsating across the valley. Rare birds, endangered birds, birds only just back from the brink of extinction, dominated by the sweet notes of the tui and bellbird. Sheer bird magic.

ABOVE In early spring tui are partial to the nectar of kowhai flowers blooming profusely on sun-warmed banks.

OPPOSITE A favoured habitat for many thriving species, forest margins occur naturally from sea level to treeline. Here a wild spaniard blooms where beech trees give way to alpine tussock, while river flats break up the forest in the Cobb Valley, Kahurangi National Park.

And a magic that tells of miracles too, because not so long ago Tiritiri Matangi consisted mainly of a trampled bit of farmland, albeit surrounded by water, where the native ecosystem held on by the most tenuous of margins. Its restoration was born from the vision of a few luminary scientists and nature lovers, and accomplished thanks to the combined efforts of a volunteer citizen group from the Auckland region, the Supporters of Tiritiri Matangi, and the Department of Conservation. Two hundred and eighty thousand native trees were planted, and alien pests removed. One rare species after another was reintroduced, from birds such as the saddleback (tieke) and kokako to the reptile tuatara. Other natives made a comeback on their own. The bellbird (called korimako or makomako in Maori) which was down to 24 individuals in 1969 (and was extinct on the nearby main-land) had surpassed 1000 by 1994, when surveys lost count. Tiri today stands both as a sample of what many parts of wild New Zealand were like long, long ago, and what they could be again if sufficient dedicated conservation efforts were mobilised.

As a result of the recent replanting, and because of its small size, Tiri offers a classic example of forest margin habitat, where birdlife is both abundant and readily visible. Wherever tall trees give way to shrubs, vines and herbs across the country, a range of species thrive in the often resulting sun-warmed microclimates, where flowers and fruit are plentiful. Besides birds, many insects and reptiles also favour this type of environment. Some breed and find shelter in the thicker vegetation while making forays into the open in search of nectar, berries, tender shoots or flying insects. Others use the opposite approach. From the northernmost point of the North Island at Cape Reinga to the wind-blasted southern shores of Stewart Island's Port Pegasus, forest margins border beaches, sand dunes, streams, rivers, lakes and estuaries, bracken fields, tussock-grass downs and reed marshes, as well as rock outcrops and scree slopes.

In the far north these margins are frequently characterised by the crimson splashes of Christmas-blooming pohutukawa trees, especially along the shorelines. Further south it is often the nikau palm, which boasts the distinction of being the southernmost palm in the world, that acts as guardian of forest edges near shorelines and rivers. Everywhere the New Zealand cabbage tree, with its uncanny resemblance to Dr. Seuss's trufula tree, stands proud, as does the New Zealand flax. While the nikau and cabbage tree supply abundant berries, flax is important to the birdlife for its copious nectar borne on stalks laden with deep tubular flowers. Perhaps the most celebrated of all the vanguard trees is the kowhai, New Zealand's national flower, with its riotous garlands of golden flowers erupting like sunshine in early spring. Preceding leaf growth and dripping in nectar, the flowers are a magnet to honeyeaters from remote riverbanks to urban gardens.

There is no doubt that forest margins are the only natural habitat in New Zealand which have been vastly expanded since humans first appeared on the scene. From the time of Maori fires long ago to modern farming and logging, a mosaic of forest borders has been created all over the land, mostly at the expense of much taller, denser vegetation. Today each woodlot, shelter-belt or small orchard, every streambank willow grove, hillside gorse thicket and remnant copse of totara trees in a cow paddock, even many a suburban shrubbery, represents to the native species some kind of miniature forest margin.

In winter, New Zealand pigeons emerge from the forest to feed on tender buds and shoots where sunshine stimulates growth.

Many small birds thrive here, and several have increased vastly in response to these opportunities. Fantails and grey warblers, whose Maori names are piwakawaka and riroriro respectively, are perfect examples. Both are adept snatchers of insects and both bring life to hedges and bushes. The fantails flit like near-weightless paper airplanes, 8 g ($^1/_3$ oz) of bird, with half of its length made up of tail feathers which flick wide like the most delicate of ladies' fans. The grey warbler is far less conspicuous to the eye; a busy, even tinier bird than the fantail, its generous two-tone trilling song emanating from leafy borders is synonymous with warm summer mornings, when the myriad white flowers of manuka and kanuka give regenerating slopes a snow-dusted look. Dramas, both natural and unnatural, are played out here too. The shining cuckoo (pipiwharauroa) relies on the grey warbler's nest to lay its eggs. When it hatches, the young cuckoo chick evicts its foster mother's own offspring and grows to independence thanks to her misguided devotion. Fantails fall prey to a more pernicious threat; introduced rats ravage large numbers of nests, sometimes even killing the dedicated mother before she is able to fly away.

A few bird species have proven themselves even more adaptable, having colonised New Zealand of their own free will as a result of the increased availability of open spaces and borders. The welcome swallow (warou), a long-time visitor, was first recorded nesting in 1956, undoubtedly taking advantage of the flying insects available along broken forest edges, from sandflies to any number of New Zealand's numerous moths, many of them diurnal and acting like butterflies. The silvereye (tauhou) appeared in large numbers at about the same time (though it was first seen in the middle of the 19th century), becoming one of this country's most successful pioneers.

Now gracing our shrubberies, vivacious flocks feed on aphids and berries, along with anything else that is both mortal and minute.

Other birds have remained more aloof. The magnificent, oversize New Zealand pigeon, known to the Maori as kereru or kukupa, readily ventures into open farmland in search of flowers, buds and berries but soon returns to the sanctity of the deep forest. More versatile is my namesake the tui, whose boisterous song, like that of so many other New Zealand birds, takes on different tones from place to place. With blue-black and purple hued plumage, filamentous white feathers woven through its nape and a white tufty throat pompom worn like a bow-tie that quivers as it sings, it was named the parson bird by early European settlers. Its notes ring out arrogantly, like very big drops of water dripping loudly into a quiet pool, interspersed with rough crackling sounds and delicate twitters so high-pitched they are barely audible. Only its smaller cousin, the red-eyed, olive-feathered bellbird surpasses the clarity of those liquid notes. In a few places, such as Tiritiri Matangi Island or parts of the Marlborough Sounds, bellbirds are still so numerous that their morning and evening songs literally suffuse the atmosphere. Each bird's voice is but a few limpid notes, delivered in slow, syncopated cadence, rising to a bell-like question mark at the end, but collectively they strike up in mesmerising harmony that mounts into delectable, pulsating crescendos.

The bellbirds and tui, like the New Zealand pigeon, do need to return to the forest to breed. Tui in particular become quite territorial in summer, when they will not range much more than a mile from the forest edge, but may commute longer distances to manmade feeding opportunities in winter. The secretive native fernbird, or matata, on the other hand, lives mainly in dense low groundcover such as bracken and rushes, nesting close to the ground, and venturing into the forest only in search of insects. The New Zealand pipit (pihoihoi) shuns trees more than all other passerines, living primarily in grasslands from coast to mountains, where it shares open spaces with one of New Zealand's waders, the banded dotterel (pohowera or tuturiwhatu).

But no matter what their preferred lifestyle, these birds suffer in common the threats that plague almost all native species. Having evolved in the complete absence of ground mammals, they remain defenceless against their attacks. From European stoat to Australian possums, introduced pests attack their nests and rob them of their food, and in many areas it is only thanks to massive and often controversial aerial poison applications that precious native species can hold their own.

In the autumn, silvereyes gorge on abundant poroporo berries, a fast-growing plant typical of damp margins.

From the beaches of Tiri to the granite outcrops of the far south, the vivid interplay of life that forest margins foster are a never-ending source of fascination and enjoyment for me. Walking along the hardy stands of nikau palms that defy the raging westerlies along the northwest coast of the South Island, near my home, I am always on the lookout for tiny fantails flashing their coquettish tails, or iridescent pigeons silently gorging on palm berries. Higher up in the heart of Kahurangi National Park, one of my greatest pleasures comes from strolling the river flats of the Cobb Valley in December. At that time glowing fields of bright yellow *Bulbinella hookeri*, or Maori onion, bedeck the sun-drenched meadows, bordered here and there with sprays of white hebe. Pipits work the open ground and bellbird song drifts from the edges of the surrounding beech forest, from whence they commute to stands of flowering flax. On a lucky day I've seen the rare New Zealand forest falcon (karearea) make a surprise attack on a flock of unsuspecting silvereyes dancing through the stiff tangles of matagouri scrub; a literal bolt out of the blue.

It is in this environment that some of the giant flightless moa used to roam until just a few centuries ago, and the matagouri's divaricating branches — a term used to describe the way the branches grow back onto themselves to create a mesh-like tangle — attest to the plant's cunning defence against the hefty vegetarian birds' massive beaks. Here, too, hunted the now-extinct Haast eagle. The largest eagle ever to exist, with a 3-m (10-ft) wingspan and tiger-size talons, it undoubtedly surveyed these open meadows from a high perch along the forest edge, swooping down onto its unsuspecting pedestrian prey. Oddly, recent DNA study suggests this giant predator was most closely related to a small, hen-sized eagle of Australia and New Guinea rather than its more distant giant cousins of the tropical rainforests.

Whether in the volcanic regions of the North Island, limestone formations of the

South Island, or granite basement rock of Stewart Island, stone outcrops create yet another interface by breaking up the forest cover. These provide an attraction to some of New Zealand's shiest natives, the reptiles. Though easy to miss, skinks are found almost everywhere, especially the mainly diurnal *Oligosoma* group, which enjoys margins where there is exposed ground for basking in sunshine. Some range quite widely through different parts of the country, while others are restricted to tiny microhabitats. All but one are live-bearers, the mother incubating her eggs inside her body and releasing about eight young who go on to grow rapidly though their first summer. Another group, the nocturnal skinks, are forest dwellers, as are the equally fascinating New Zealand geckoes.

No doubt the most extraordinary reptile of all is the tuatara, oft referred to as a living fossil. Indeed its history goes back an incredible 220 million years or more, and to look at its angular face is to stare into a past so distant it is impossible to imagine. Here is a creature that saw the rise and fall of dinosaurs and watched the emergence of birds on the planet, all the while undergoing few changes in its own make-up. Its ancestors were once widespread round the world, but only in New Zealand has it stayed with us to the present — barely. It's life is as slow and patient as its history. With a preferred body temperature around 10 to 20°C (50–68°F), the tuatara is active mainly at night. It may not breed until it is 10 to 15 years old, and then the female will lay a small clutch of eggs a year after mating. Buried in the cool forest floor, these then take an incredible 12 to 15 months to hatch. Although no-one has followed an individual tuatara for long enough to be sure, it is believed a 60-cm (2-ft) male may be well over 60 years old.

It is probable that originally tuatara favoured forest margins everywhere in New Zealand, as their bones show they were once widespread. But today, they survive only on a small number of offshore islands where rats have not invaded to devastate their young. On Stephens Island, they may number as many as 2,000 per ha (800 per acre), with a total population for the island around 30,000 to 50,000. With its razor-sharp wedge-shaped carnivore teeth and its third eye embedded on the top of its skull, the tuatara is indeed a prehistoric animal befitting of our awe and wonder — the more chilling to realise how close we have come to losing it.

LEFT The shape of its beak matched to the curve of the flax blossom, a bellbird works the nectar-rich stalks like traplines, helping to pollinate the plants.

ABOVE Active volcanoes create natural forest edges, where plants such as this lancewood sapling, take advantage of the sunlight. The lancewood is unusual in that the long, downward-pointing juvenile foliage converts to short, upright leaves as it matures; a trait called heteroblasty.

LEFT AND RIGHT One of the oldest New Zealand plants, the common flax, or harakeke, is a lily with tubular, nectar-filled flowers irresistible to tui and many other birds. It grows along streams, swamps and treelines at all elevations, and is replaced by a smaller species on exposed seashores and mountains.

ABOVE Primarily an insect eater, a saddleback emerges from the undergrowth to exploit the energy boost it can gain from abundant pohutukawa nectar.

LEFT Fluffing up its resplendent feathers, a dominant male tui stands guard over a rich nectar source, keeping other birds away from the precious food supply.

LEFT Where nectar-bearing plants are plentiful, the delectable song of the bellbird rings loudly. Females, such as the one pictured here, sing as actively as males.

BELOW AND RIGHT Belonging to the myrtle family, gnarly old pohutukawa trees bedeck northern coastal areas with brilliant Christmas-time blooms, whose nectar is a key food source for nesting birds. Like rata in the south, the trees are suffering from land clearance and possum defoliation.

FOLLOWING PAGES With brush-tipped tongue and agile behaviour, tui feast on tubular kowhai blossoms, carrying pollen from tree to tree on their foreheads. Largest of the honeyeaters, they may fly 30 km (18 miles) from their forest home to a secure source of nectar, which makes up 80 percent of their diet, besides insects and seeds. Males are larger than females, though both sing loudly. Behaviour includes dominance displays and sun bathing.

OPPOSITE Nikau palms and flax line the Kohaihai River mouth in Kahurangi National Park.

ABOVE LEFT New Zealand pigeons depend heavily on nikau fruit to feed their chicks, who can reach full adult weight (over half a kilo) in less than four weeks on such nutritious food. Since the extinction of the moa, pigeons are the only fruit eater large enough to disperse such seeds long distances.

ABOVE RIGHT AND RIGHT Of some 1,100 species worldwide, the nikau is the southernmost of all palm species. Growing to 9 m (30 ft) tall, dense stands line the western shores of Kahurangi National Park.

ABOVE LEFT AND RIGHT Along with flax, *Myrsine* and *Coprosma*, the fast-growing, wind-resistant kanuka and manuka trees represent the *avant garde* of the forest, their compact canopies joining like a mosaic and offering shelter to any wildlife that favours the borderlands.

RIGHT A tuatara basks in dappled sun to warm itself. This ancient reptile uses sharp cutting teeth and a special shearing jaw motion to prey on birds, geckoes and insects.

BELOW Some of New Zealand's enigmatic stick insect species comprise only females, which reproduce asexually by hatching unfertilised eggs, whereas others breed more conventionally and hybridise readily.

ABOVE Their rigid fronds able to resist buffeting westerly gales, nikau palms dominate the scrub forest at the top of the South Island, where tuatara would have abounded before the introduction of predatory mammals. Beyond stretches Farewell Spit, 30 km (18 miles) of shifting sand dunes and 10,000 ha (25,000 acres) of ecologically-rich tidal flats.

LEFT The only remaining representatives of a reptile group that thrived during the dinosaur era, tuatara are extremely slow growing. Their eggs take 12 to 15 months to hatch and they reach full size — for males up to 60 cm (2 ft) in length and 1 kg (just over 2 lb) in weight — in 25 to 35 years. Preferring broken forest environments, and possibly living for as long as a century, today two species survive only on islands free of introduced predators.

BELOW The common blue butterfly, along with many diurnal moths, is one of a bevy of native insects that frequent meadows replete with introduced weeds.

ABOVE (LEFT AND RIGHT) AND RIGHT In the past, welcome swallows were only visitors to New Zealand, but with the expansion of open ground they started nesting in 1956 and are now a feature of farmland and forest edges where flying insects abound. Soon after leaving the nest, chicks are still fed assiduously by their parents.

Silvereyes arrived naturally from Australia and made their New Zealand debut in the middle of the 19th century, taking advantage of expanded opportunities in the wake of shrinking forests and vanishing competition from endemic birds like brown creepers and whiteheads. They feed busily on anything from aphids and tiny insects to nectar, fruits and berries, and form large flocks in winter.

LEFT Beech forests offer sun-warmed pockets along their edges, where even heavy winter frosts soon burn off.

ABOVE A springtime blaze of Maori onions, or *Bulbinella*, is the hallmark of mosaic alpine meadows that break up the forests of the Cobb Valley.

BELOW New Zealand pipits seek fellfields and open ground beyond the forest edge.

ABOVE Rocky outcrops, such as Baldy Head overlooking Port Pegasus in southern Stewart Island, compress the forest into a narrow coastal band. The granite extrusions, where only stunted plants grow, were once deeply bedded magma which cooled far below the surface, representing some of the oldest visible rocks.

LEFT Cave wetas, one of many species of this ancient group of insects, deck the walls of their underground hideaway. RIGHT In the north of the South Island, the spectacular karst limestone formations produce a varied habitat riddled with caves and underground rivers.

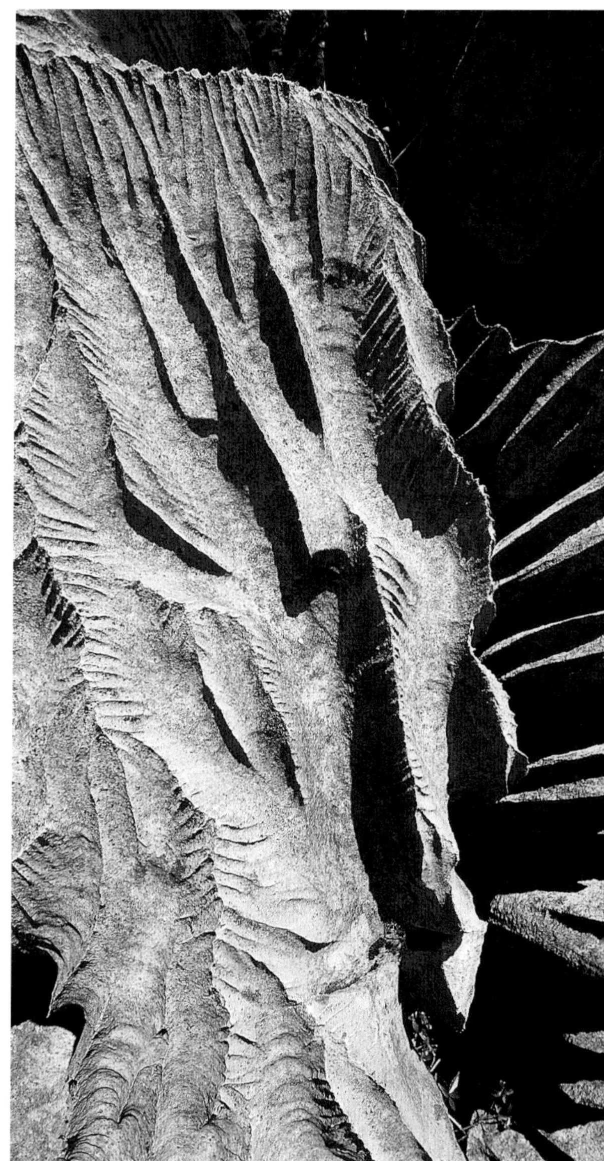

ABOVE LEFT The Anatoki River flows through the ecologically-rich and varied Kahurangi National Park lowlands, where lush stands of podocarp and broadleaf forests give way to grassy river flats.
ABOVE RIGHT Feeding on *Coprosma* berries, a New Zealand pigeon helps to spread seeds across broad valleys.
LEFT Living in reed beds and ferny bogs, the secretive fernbird flies only very short distances, remaining close to the ground.
RIGHT Fantails chase flying insects along the forest edge.

ABOVE A saddleback seeks out insects in the shaggy crown of a cabbage tree.
LEFT A kokako feasts on cabbage tree berries in summer.
RIGHT The New Zealand cabbage tree is referred to alternately as the largest lily or the largest agave. Reaching 3 m (10 ft) in diameter and a height of 12 m (40 ft), it grows a large, carrotlike, starchy tap root below ground. Natural survivors, they often stand where more delicate forest has succumbed to fire or farming and attract many birds to their bushy fronds.

2
HIDDEN REALM OF THE KIWI
LIFE IN THE MOSSY UNDERSTORY

There is something tinglingly primordial about being deep in New Zealand's forest at night. On a moonless, windless evening few places on this planet, except underground, are as totally enveloped in inky darkness, or resonate with such a deep, hushed silence. No starlight penetrates the tiered canopy, and the overarching stillness speaks of myriad hidden lives busily doing what forest creatures do, unseen and unheard. Even though the benign New Zealand fauna holds no fierce carnivores, venomous snakes or killer insects — on the human scale that is — the feeling of unseen presence harks

ABOVE *Blechnum*, or crown, ferns bedeck the damp forest floor on rat-free Ulva Island Sanctuary, a haven for ground-dwelling birds.

LEFT Small but bold, a male Okarito brown kiwi or rowi, a newly identified distinct species, ventures out of his burrow in daylight to escort intruders off his territory deep in the coastal forest.

back to our deepest instincts, sharpening all of my senses to hair-trigger acuity.

Only once my ears adjust to the stillness do I pick up the faintest telltales of activity. The soft interrogation of a morepork (ruru), a tiny forest owl, drifts from somewhere across the valley. And from the leaf litter around me emanates an almost imperceptible *frisson* of primitive gnashing jaws, for the night armies are at work.

Giant weta, obese flightless grasshoppers with threatening looks but gentle personalities, have come out of hiding to chew on plant matter. Numbering some 70 species, which come in many shapes and forms, the largest represent some of the heaviest insects in the world, with stout bodies up to 9 cm (3½ in) that approach 80 g (3 oz) in weight. Weta are found only in New Zealand and its offshore islands and are living representatives of an insect group dating back 200 million years, living and acting much like mice and rats, whose ecological place they occupy.

An equally startling impersonation of the role of forest-floor rodents is played out by one of New Zealand's only two land mammals, the short-tailed bat. Although it may catch insects on the wing by using biosonar, or echolocation, like many other bats, it also spends much of its time scrambling nimbly on the ground, seeking out a variety of high-energy foods from crawling insects to fruits and nectar. One of the most amazing discoveries about this unique bat is its intimate association with an obscure parasitic plant, *Dactylanthus taylorii*. Growing underground by invading the root system of trees, *Dactylanthus* produces drab though nectar-laden flowers. Barely emerging from the substrate, these flowers rely on the pedestrian bats for pollination.

The newly discovered Maud Island frog is a primitive creature with no voice and no eardrums, whose eggs hatch into froglets instead of tadpoles.

New Zealand's rare and endangered frogs, little changed since time immemorial, also live on the forest floor. Shunning standing water, they reproduce without passing through a tadpole stage. Instead, the male broods the clutch of large eggs, from which tiny froglets with vestigial tails hatch to climb onto his back and complete their development. Just three tiny species under 5 cm (2 in) long survive in minute pockets of forest, where pernicious chytrid fungi, which infect frogs ultimately causing death, have recently placed their survival in question.

Geckoes and skinks are the only reptile representatives in New Zealand besides the extraordinary tuatara. They come in a bewildering array of species — at least 70 and still counting, as more are being discovered occupying widely differing habitats — all of whom blend superbly with the dappled patterns of leaf litter, tree bark, mosses and ferns. No doubt this was a valuable survival trait when the bizarre, extinct lizard-eating adzebill used to prowl the forest realm, festooned as it is with equally otherworldly plants. Here live chlorophyll-free orchids, which sustain themselves by sharing the nutrients that underground fungi rob from tree roots. Meat-like stinkhorns splay putrid-smelling fruiting bodies designed to be irresistible to flies, and tree trunks are coated with honeydew moulds upon which bellbirds rely for sugar. There are also basket fungus, spine fungus, coral fungus and jelly fungus, and a dazzling collection of red, yellow, white, blue and green gill-fungi.

In the daytime, dancing among filmy ferns and umbrella mosses, busy New Zealand robins (toutouwai) — one race for each of the North and South Islands — flush insects from the leaf litter by vibrating one foot on the ground, then watching for movement. Saddlebacks or tieke, also with distinctive north and south forms, use their dagger-straight beaks to extract their prey from rotting logs and the scaly stumps of giant tree ferns.

Perhaps the most amazing little bird of the deep forest is the diminutive rifleman (titipounamu) and its recently extinct relatives, the bush wrens. Not true wrens at all, these minute, long-legged, tail-less New Zealand birds are regarded as the oldest existing representatives of the vast group of passerines — or perching birds — going back at least 80 million years, with no other living relatives in the world. Several species were flightless and are known only from their bone remains, apparently living entirely on the ground and behaving amazingly like the mice that were absent from the New Zealand fauna. Not surprisingly, the arrival of the first rodents caused upheaval early on, and several species of wren disappeared very soon after Pacific rats were brought to New Zealand by early Maori. The last truly flightless species to vanish lived on Stephens Island off the north tip of the South Island. For a sadly brief period at the end of the 19th century, it became better known to the lighthouse keeper's cat than to science.

The more widespread bush wren, which flew very little, was last seen about a half century ago. Likewise, its close relative, Stead's bush wren, lived only on Big South Cape Island in the far south of the country, where it too vanished abruptly when ship rats were accidentally introduced in the early 1960s. Only the diminutive rifleman, just 8 cm (3 in) long and weighing less than 7 g (½ oz), is still fairly widespread, especially in virgin beech forests. Clearly, it owes its survival to being more arboreal than all of its relatives. While pairs can often be seen scurrying about in the darkest nooks and crannies of the forest floor, they are equally at home flitting up and down the trunks and gnarled boughs of mossy trees. Another relative, the rock wren, leads a ground-loving existense in the high mountains above treeline.

Strange plants and unique vertebrates notwithstanding, the New Zealand forest floor, especially at night, truly belongs to the boneless guild, though not all members of this rich invertebrate society are genteel in nature. Indeed here is a relict world from the age of dinosaurs where many a night stalker is every bit as fierce as *Tyrannosaurus rex* ever was, albeit on a Lilliputian scale. Centipedes up to 20 cm (8 in) long tackle prey, including geckoes, using venomous bites. Over 1,000 species of spiders lurk, 95 percent of them endemic, including the cave spider with a leg-span of 15 cm (6 in) — with perhaps twice as many species awaiting discovery.

As with the weta and centipedes, gigantism is a recurring theme among the nocturnal forest floor community, of which many groups are represented by a staggering numbers of species. Working the ground cover, several hundred types of millipedes recycle the leaf litter, including one 10-cm (4-in) monster. But just beneath the surface, this exploit pales against the recyclers of all time, the earthworms. The Goliath of New Zealand's roughly 200 species measures an incredible 1 m, or 40 in, in length and 1 cm (⅜ in) in diameter. Even springtails — primitive, wingless jumping insects usually barely visible to the naked eye and occurring everywhere in the world in phenomenal density — may grow to 14 mm (½ in) or more in New Zealand.

Of the more than 1,000 New Zealand land snails, the carnivorous giant *Powelliphanta* comes in dozens of beautifully ornate species, some up to 9 cm (8½ in) across.

Land snails are no exception to this gigantic, meat-eating trend. They, too, show prolific diversity with over 1,000 species recorded, many of them restricted to tiny pockets of habitat or endemic to a single hill, acting as an island in the varied landscape. The *Powelliphanta* group are fiercely predatory and include fist-sized giants that hunt earthworms, slugs and other prey at night. With over 70 species and subspecies in total, they are thought to belong to the world's oldest carnivorous snails, once widespread across Gondwana. Their beautifully whorled shells grow to 9 cm (3 ½ in) across, and their heft has been compared to that of a female tui at over 90 g (3 oz). What's more, they may live for over 20 years and don't start breeding until five or six years old, laying small clutches of hard-shelled, pinkish-white eggs, some 12 to 14 mm (over ½ in) in length.

Powelliphanta snails come in a delightful array of colours and striped patterns, ranging from rich caramel gold to cinnamon brown and dark coffee. The northwest part of the South Island is the stronghold for a bewildering concentration of forms, many of them living in tiny microhabitats scattered from sea level to above the treeline. Sadly, over half are threatened, none more so than the one discovered in 1996 on Mount Augustus near the West Coast, where rainfall can reach 6 m (20 ft) per year. The site of an open cast coal mine, the last remnant of its 1,010-m (3,300-ft) high home is under threat, in spite of intense protestations by conservation groups and the Department of Conservation. To make matters worse, feral pigs, rats and even introduced blackbirds, thrushes and hedgehogs all see land snails, whether carnivorous or not, as delicacies.

The kauri snails, or *Paryphanta*, are the far northern equivalents of the *Powelliphanta*. Nearly as large and just as predatory as their southern cousins, they distinguish themselves in their ability to travel remarkable distances — not only on the ground but also high up tall trees — in search of prey. Also at the northern tip of the country lives a small group of very different giants, the huge flax snails, or *Placostylus*. Sedentary vegetarians with fluted, nearly 12-cm (5-in) long shells, all four species are critically endangered by the loss of their preferred habitat to farming, and predation by introduced mammals.

Not to be outdone by the array of oversize forest-floor killer invertebrates, the tiny New Zealand glow worms have turned the darkness of night to their advantage by dazzling their prey with light. On moist banks and in caves, glow worms, the larvae of a small gnat, weave net-like traps made of thin strands of sticky mucous to ensnare their victims. Living in colonies, each night they switch on entire constellations of bluish 'stars', using bioluminescent abdomens to lure other small insects to their death. After many months of worm-like larval existence, the adult gnat dies within days of hatching, its reproduction turning into a race against the clock.

Of the weird and wondrous New Zealand retinue of blood-thirsty predators, the prize must go to the most enigmatic of them all, the velvet worm, or *Peripatus*. A relic creature tracing its ancestry back a mind-numbing 500 million years, it is believed to somehow span the gap between primitive worms and segmented insects. With 15 pairs of legs resembling those of a caterpillar, and tiny stalked eyes like a snail's, its soft looks provide no hint of its bloodcurdling hunting habits. When it senses its prey in the damp leaf litter it first 'shoots' it with sticky slime to immobilise it. Thus entrapped, the live victim is sliced open with sharp mouthparts and injected with digestive juices to be sucked out as a slurry.

It is in this strange, worm-eat-worm underworld, during the dark of night, that I first met New Zealand's most celebrated icon of all, the kiwi. Large numbers of outdoor enthusiasts sensibly walk the famed Heaphy Track through Kahurangi National Park in the daytime, when nikau palms and pigeons, beech forest and bellbirds saturate the senses. But meeting the night shift brings a very different excitement. Pondering the secretive interplay of life under the mantle of darkness, nothing prepared me for the abrupt approach of heavy, assertive footfalls crunching down on the leaf litter. Suddenly there she stood, a female great spotted kiwi barely outlined in the red glow of my screened torchlight, 45 cm (1½ ft) of bird standing proud and stock still in a defiant, upright posture, bill tucked to her chest and heavy foot cocked and ready to strike at the first hint of an intruder into her territory. As I held my breath she ducked into the undergrowth and circled around me, inspecting my presence by sniffing the air audibly like a mammal might, before heading on down the slope. A moment later, the frosty night air was stabbed by a bloodcurdling volley of hoarse, questioning screams, echoed within a few instants by an equally potent yet shriller response from her smaller mate (females are nearly half as heavy again as males) down the valley across the Saxton River — 'half whistle, half scream' as described by Sir Walter Buller, New Zealand's early naturalist and pioneering explorer.

By walking the Heaphy Track at night, when all other trampers are safely tucked into their sleeping bags in well-spaced huts, I have had many encounters with the great spotted kiwi; at over 3.3 kg (7 lb), the largest of the various forms once found across New Zealand. Distantly related to ostriches and the extinct moa, kiwi are so unlike all other birds that they have been dubbed an 'honorary mammal' in recognition of their un-birdlike features. With small eyes, long sensory whiskers, an acute sense of smell and nostrils placed at the tip of their somewhat flexible bills, they indeed have evolved a lifestyle akin to mammals. Their vestigial wings are but minute stumps, while they lope about on springy, muscular legs that make up about a third of their weight, nightly patrolling territories of up to 60 ha (150 acres), where no other kiwi but long-standing mates are allowed to trespass.

With sensitive whiskers like a mouse, a North Island brown kiwi peers through the entrance of its burrow.

Kiwis are mammal-like in a whole range of ways. Not only do they possess a doglike sense of smell and catlike sensitive whiskers, they live in burrows like rabbits. Their bones are marrow-filled rather than hollow, and their body temperature is several degrees lower than most birds, about the same as ours. They also lay the largest eggs in relation to body of any bird. The egg can represent up to a quarter of the female's own weight, perhaps explaining why she is considerably larger than her mate. But in many cases he alone incubates the egg for an incredible 2 to 2½ months, hatching out a mini-kiwi ready to start off on its own, with no further help from either parent.

For a long time only three species of kiwi were recognised, set apart in their own separate order the Apterygiformes. But recent studies based on DNA analysis have demonstrated that they are more diverse genetically than ever imagined. Currently, five species have been accepted: the great spotted kiwi, or roa; the little spotted kiwi, the North Island brown, the southern tokoeka (a form of brown kiwi on South and Stewart Islands) and the rowi, or Okarito brown kiwi, restricted to a small section of forest in Westland National Park. Two more distinct races are also under scrutiny: the Haast tokoeka to the south of the rowi, and the Fiordland region version of the southern tokoeka. The one thing all kiwis share in common is troubled times due to predation and competition with introduced mammals, particularly the lithe and ferocious stoats brought from Europe in a futile attempt to control proliferating plagues of rabbits released by earlier settlers.

Deep in the sodden Okarito forest I was able to gain an intimate understanding of the rowi's quirky way of life. Intensely territorial, this kiwi sometimes even emerges from its burrow in broad daylight just to escort you off its turf. Setting up camp near a nest where a radio-carrying chick was making its first forays out of the parental burrow, unescorted and untutored and already independent-minded at two weeks of age, taught me just how wondrously odd this bird really is.

Further south, on Stewart Island, I saw another side of the kiwi's personality when I spent nearly an entire night in the company of a large female southern tokoeka busily feeding on sand hoppers thriving in the strandline of kelp washed up on a beach. Oblivious to my presence, she poked and prodded, sniffed and sneezed, deftly thrusting her probing beak down to the hilt into the sand. Eyes closed, she never paused, gobbling the wriggling, hopping prey with her expert combination of keen smell, soft-tipped, flexible beak and touch-sensitive vibrissae (hairlike 'whiskers' around the base of the beak). After five hours of feeding she moved to a small stream to drink, then met her mate near the edge of the forest where they growled, chased and pranced for some time before vanishing back into the obscurity of the thickets. It was an unforgettable moment shared with one of the most eccentric birds on the planet.

These memorable experiences with New Zealand's strange flightless world were surpassed only with a rare visit to Codfish or Whenua Hou Island off Stewart Island's windswept west coast, where three species of penguin — yellow-eyed, little blue and Fiordland crested — also venture into the cool undergrowth to nest. Tucked away in this pristine wilderness forest, the crown jewel in New Zealand's extraordinary trove of living biological treasures is making a steady comeback from the very edge of oblivion. The world's heaviest and strangest parrot, the nocturnal kakapo, has been snatched away literally at the eleventh hour from the jaws of extinction.

Much has been written about this bizarre avian anachronism: a parrot who, unable to fly, scrabbles about in the thickest of mossy fern understory; who booms through the night to attract a mate and hides in the day thanks to its perfectly camouflaged feather design; whose plumage is scented like a tropical fruit; and whose beak is designed to mush berries and seeds on the forest floor. Its facial expression resembles that of an owl thanks to feathery disks funnelling sound towards its ultra sensitive ears, while its bristling hairy 'whiskers' enable it to feel its way where lack of light limits vision.

No amount of literature could prepare me for the sight of this utterly enigmatic creature emerging so quietly from amongst the fern fronds in the midnight hours that not a leaf trembled, its oversized parrot feet placed so deliberately and stealthily that not a twig scraped on the forest floor. Through trial and error, failure and breakthroughs, and above all dogged determination spanning the last three decades or more, team after team of researchers, wildlife officers, Department of Conservation staff, biologists, vets, managers, sponsors and volunteers, united under the banner of the Kakapo Recovery Programme, have finally turned the tide of doom for this icon parrot.

The kakapo's disaster began with human invasion of its habitat centuries ago, and its future now resides in an intensively managed, steadily growing population centred on Codfish Island. Every single kakapo in the wild wears a computer chip under its skin and a small backpack with a radio transmitter by which its whereabouts are closely tracked and its health monitored. Adults who appear stressed are offered food supplements in regularly stocked hoppers. At painfully rare intervals, when the rimu trees produce an over abundance of fruit, breeding fervour may spread for a season, causing males to boom their lovesong from strategically selected sites to entice females into mating. The chicks that ensue are treated by the island managers as more precious than gold dust — and so they should be. Slowly but surely kakapo numbers are climbing from an all-time low of 51 in 1995 to 86 in existence at the time of writing. Watching last year's rambunctious hand-reared youngsters discover the earnest pleasures of tasting wild berries for the first time, I realised that saving the kakapo is not just a matter of national responsibility or pride for all New Zealanders, but a question of saving our own souls as the global custodians of all things wild and precious.

The world's heaviest parrot, the nocturnal kakapo, is also one of the rarest, making a slow comeback through intensive conservation efforts.

ABOVE AND BELOW With large eyes and slender legs, a South Island robin hunts for insects among kidney ferns and umbrella moss by vibrating one foot in the leaf litter, then watching for movement while remaining perfectly still.

RIGHT The largest of all tree ferns, the giant mamaku grows to 20 m (65 ft), creating a filigreed canopy shading the dark, moist forest floor.

BELOW Snuffling among fallen logs, a North Island brown kiwi uses its keen sense of smell to detect insects and worms under the forest floor debris.

RIGHT Beneath a luxuriant understory of damp ferns and mossy trunks, a microcosm of forest floor organisms live largely unseen.

BELOW Twenty-seven species of filmy ferns, including the kidney fern *Trichomanes* shown here, cover the forest floor and coat tree trunks, where shafts of sunlight make scanty incursions.

LEFT Almost unchanged for the last 200 million years, the giant weta is among about 70 species of flightless grasshoppers, some up to 10 cm (4 in) long, that evolved to live like rodents in the absence of ground mammals.

LEFT A particularly colourful form of giant carnivorous *Powelliphanta* snail is confined to the beech forest of Takaka Hill and Abel Tasman National Park.

ABOVE AND BELOW RIGHT Caves and overhangs light up with nightly constellations of glow worms. The larvae of a tiny gnat, they spin sticky netlike snares to trap even tinier insects attracted to their light displays. Known as bioluminescence, a complex enzyme-driven chemical reaction combines luciferin, luciferase, oxygen and other components to convert energy into light. The larvae live for 6–9 months, whereas the adults, who have no mouths, breed and die within a few days.

BELOW LEFT The damp forest floor is home to a bewildering variety of fungi.

With new clues emerging from DNA studies, five species of kiwi are currently recognised. Two more distinct races are also under scrutiny. CLOCKWISE FROM TOP LEFT North Island brown kiwi, *Apteryx mantelli*; great spotted kiwi or roa, *A. haasti*; rare Haast tokoeka whose exact status is not yet defined, confined to a small area of Westland; Okarito brown kiwi or rowi, *A. rowii*; Stewart Island southern tokoeka, *A. australis*; little spotted kiwi, *A. owenii*, now restricted to offshore islands. Another form of southern tokoeka awaits classification in Fiordland.

ABOVE About 45 cm (1½ ft) long, the great spotted kiwi is the largest kiwi species. It lives mainly in Kahurangi National Park from seashore to tussock highlands, where pairs nightly patrol territories of up 40 ha (100 acres).

RIGHT Sometimes active in daytime, it constructs deep burrows or uses multi-entrance galleries beneath boulders and fallen logs.

ABOVE RIGHT AND FAR RIGHT Worms and insects are pulled deftly from the leaf litter.

ABOVE Sandwiched between the sea and Lake Mapourika, the Okarito forest in Westland is the only known home of the recently recognised rowi, or Okarito brown kiwi.

RIGHT Leading a pedestrian life in a twilight world, a male rowi prowls beneath the umbrella fronds of *Blechnum* ferns.

FOLLOWING PAGES Emerging from the forest under the cover of darkness, a female southern tokoeka on Stewart Island's Ocean Beach works the wet strandline, using her acute sense of smell to detect buried sand hoppers. With nostrils at the tip of her flexible bill, she thrusts and probes purposefully, also snapping up her quarry in mid-air with the help of sensitive whiskers. After feeding for five hours, she drinks from a small stream before returning to the forest.

ABOVE LEFT AND RIGHT Weighing up to 3.4 kg (8 lb), the kakapo is the world's heaviest parrot, scurrying through the understory and even scrambling up trees using its strong beak and feet.

LEFT AND RIGHT A night's foray includes finding *Astelia* berries and extracting the sweet juice from grass stalks on the forest floor.

LEFT (TOP TO BOTTOM) Kakapo feeding on manuka seed pods, supplejack and mingimingi berries.

RIGHT Clambering through scrub and up vines, a kakapo uses a flat plate inside its beak to crush juicy berries and seeds.

NEXT PAGE Flightless and nocturnal the kakapo relies on a keen smell, sensitive hairlike feathers around its beak, and its owllike facial disks to focus sound. With just 86 birds alive today, the species is making a slow comeback thanks to intensive management on predator-free islands such as Codfish/Whenua Hou.

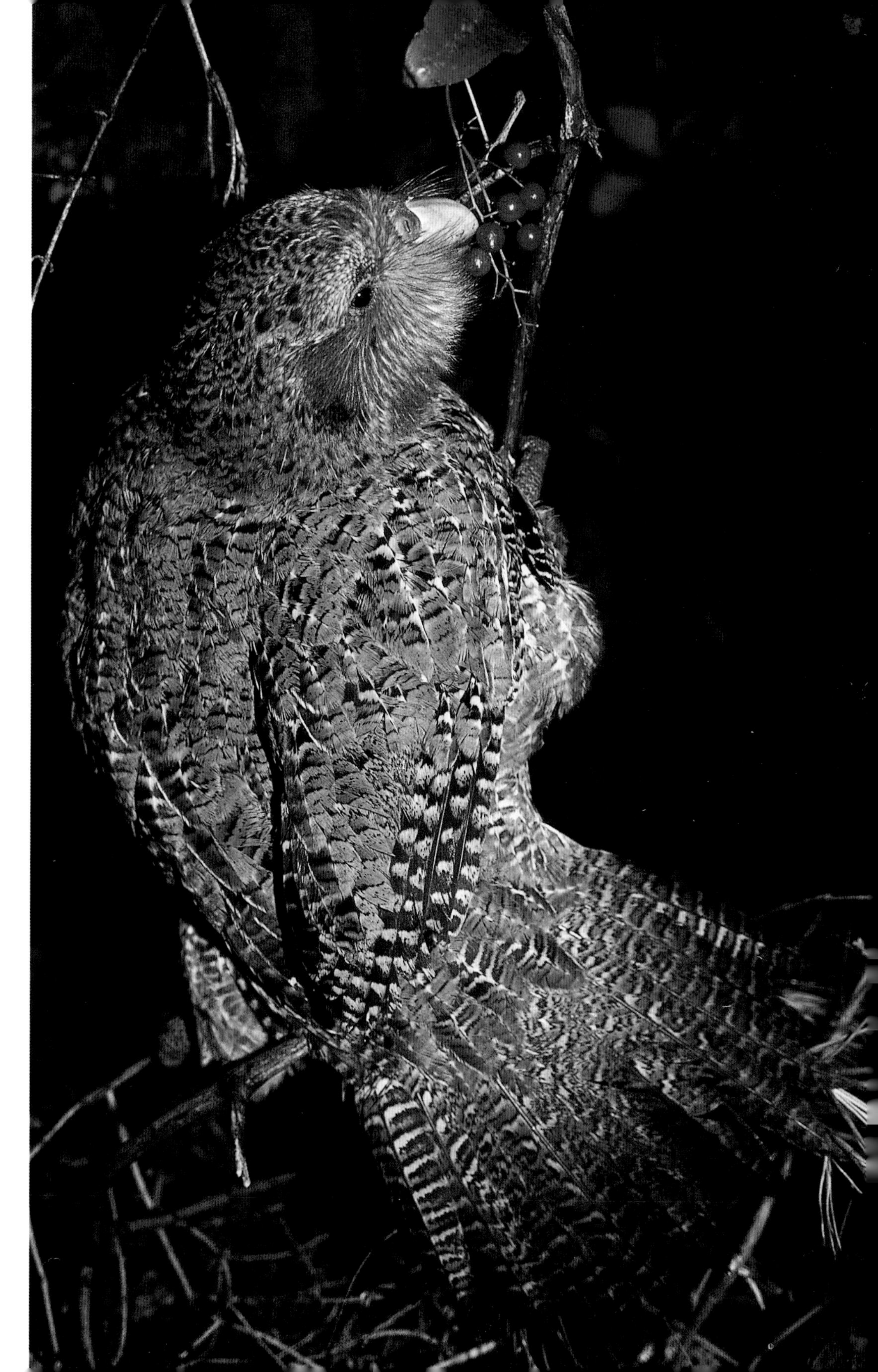

Three species of penguins use the dark recesses of the forest to nest well away from the shoreline.

RIGHT Standing just 40 cm (16 in) tall, a little blue penguin — the world's smallest — commutes under the cover of night to its nest hidden in fern thickets in Golden Bay, South Island.

FAR RIGHT Yellow-eyed penguins prefer the southeastern coast and subantarctic islands, nesting under coastal scrub, where a pair exchanges loud greetings at dusk.

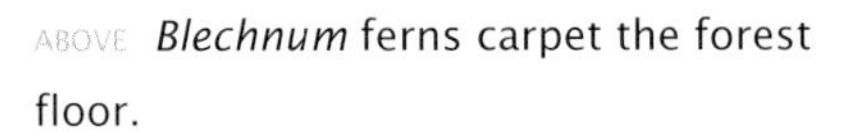

ABOVE *Blechnum* ferns carpet the forest floor.

RIGHT A small colony of Fiordland crested penguins, perhaps the world's rarest, gathers deep in a wet gully on Codfish Island.

Never infested with rodents, Maud Island, tucked away in the Marlborough Sounds, is a haven for ground-dwelling wildlife.

ABOVE A Maud Island frog sits motionless on the forest floor waiting for insects to pass within range. Emerging from deep within boulder slopes on rainy nights, they shun standing water, the male carrying the eggs on its back.

RIGHT AND FAR RIGHT The dainty, nocturnal common forest gecko, its camouflaged colours varying with the substrate, thrives where predators are absent. More and more diverse gecko species are being discovered in every region of the country.

RIGHT A pregnant female tunnelweb spider in its silky lair also benefits from the absence of rodents on Maud Island.

OPPOSITE A solid wall of podocarp forest dominated by swamp-loving kahikatea trees borders Lake Mapourika in Westland, sheltering a microcosm of little-known life in its dark, dank interior.

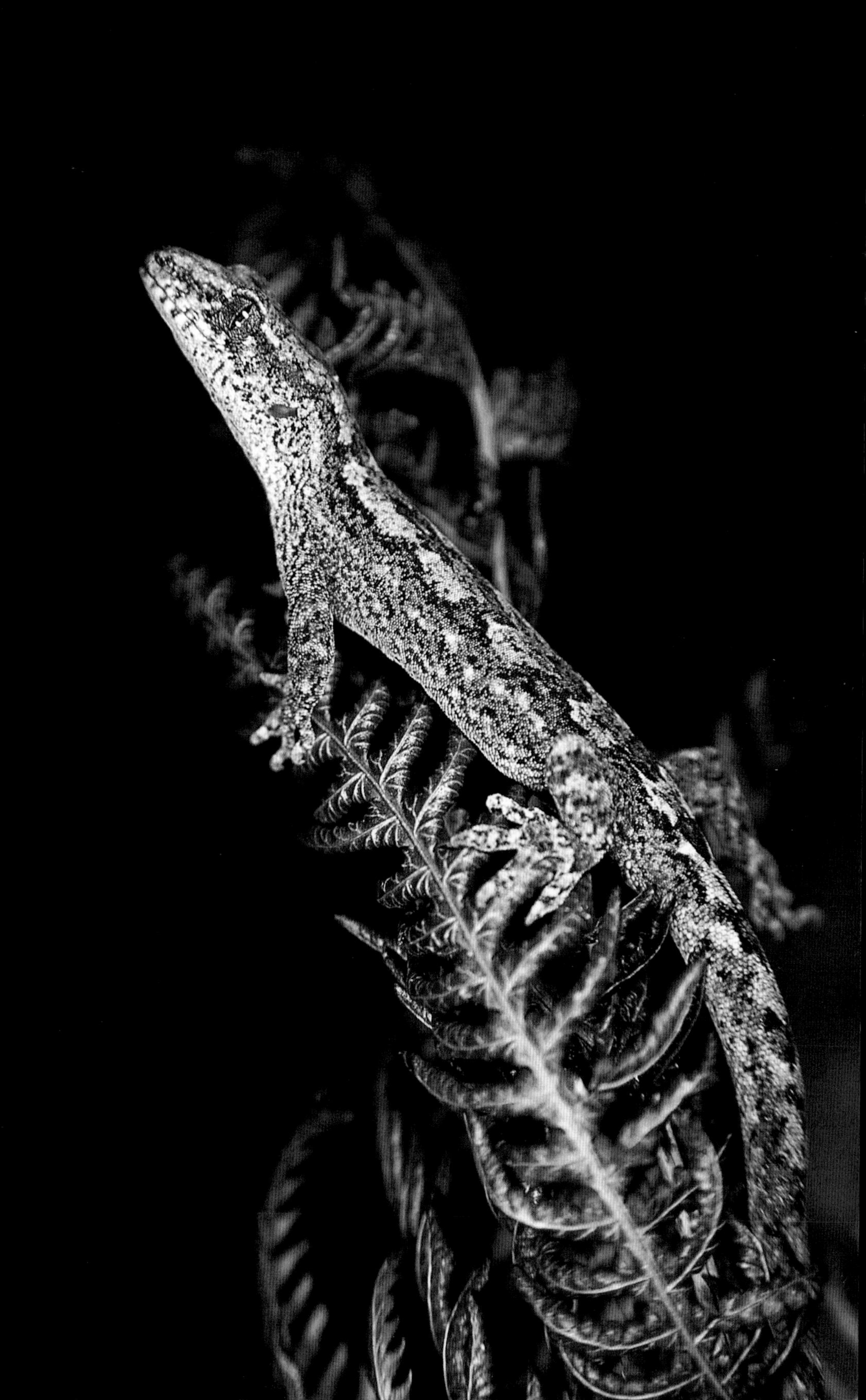

3

WINGS ACROSS GONDWANA FORESTS

UNTAMED SOUTH ISLAND

Airy is the first word that comes to mind when I think of the South Island's splendid beech forests. Unlike the helter-skelter layout of wet lowland forests, the southern beeches, *Nothofagus* species, seem to grow in an orderly and organised fashion, as if some long-term plan was the underlying factor in their design. Though often dense, their compact branches spread out in tiered rather than extravagant layers, creating a breezy effect that is somehow pleasing to the eye. A springy cushion of golden leaflets spreads silently underfoot, and carpets of moss envelop the ground in the wetter places.

ABOVE A male New Zealand falcon, or karearea, swoops down from its perch, capable of diving onto its unsuspecting prey at speeds up to 180 kph (over 110 mph).

LEFT Harking back to ancient times, some of the wildest forests in the world stretch unbroken from the shores of the Tasman Sea to alpine snows in Westland National Park; looking east from the Okarito Trig Point.

The New Zealand beech trees, of which there are just four species, are mirrored by nearly identical forms in distant Patagonia. They grow slowly and speak of ancient times, confirming that New Zealand and South America once shared common ground untold millions of years ago, when both were still attached to Antarctica. Although southern beeches occur in parts of the North Island, these forests are the hallmark of the cooler, drier, higher elevations of the South Island.

At home in the dappled mid-canopy, like a fleeting, high-strung but self-assured spirit, a New Zealand falcon, or karearea, swoops in amid the dark boles and lands on a jagged snag. 'Kek-kek-kek-kek-kek-kek-kek', the lithe male, weighing just 300 g (10 oz), proclaims his right over his domain, while his larger mate, about two-thirds again his weight, is nestled on the ground in a bed of tiny dry leaves. She has made her nest scrape under the shelter of an overhanging boulder perched gingerly on the steep forested slope.

They are assertive, agile little raptors, adept at hunting down small forest birds on the wing. And they are also survivors among the top avian predators that once ruled the New Zealand ecosystem. Not so many centuries ago the now extinct Eyles's harrier, a chunky raptor weighing perhaps 3 kg (6 or 7 lb), was apparently a wait-and-pounce forest predator preying on any kiwi or kakapo — no doubt common birds back then — careless enough to make a daylight appearance. Likewise the overpowering Haast eagle, whose talon scrapes have left evidence that it was capable of felling a moa weighing 250 kg (550 lb) or more, patrolled the forests. This, the largest eagle ever to exist, was once the ruler of the South Island — its only home — and is believed to have survived until just a few centuries ago.

An important source of nectar for beech forest birds, parasitic mistletoe is threatened by introduced possums.

With its agility and small size, the New Zealand falcon, a forest-loving relative of the worldwide peregrine, has survived its larger cousins primarily in the South Island. Deep in the splendid beech forest of Kahurangi National Park, Mark and I followed the nesting pair through their daily tribulations. No one and nothing moving within several hundred feet of their ground-level eyrie could escape detection, incurring immediate and fearsome dive-bombing wrath. However, by gradually introducing an unobtrusive, camouflaged photographic blind, positioning it progressively closer over several weeks and always using one of us as a decoy when making our furtive visits, we were able to peak unnoticed into their private lives.

The male spent most of his time hunting while his mate spent her days guarding and warming the two helpless, fluffy chicks. Highly territorial, with boundaries known only to neighbouring falcons, their domain clearly was patently productive. Every few hours he would return, loudly announcing another successful hunt. His mate flew up to meet him on a high perch to accept the proffered prey, a bedraggled feather mop clutched tightly in needle-sharp talons. Back at the nest, she placed small morsels tenderly in tiny beaks with all the dedication of a doting mother. Bellbirds, tomtits and silvereyes, plus song thrushes and blackbirds, all figured on the menu, though birds as large as the New Zealand pigeon may be tackled on occasion.

Thriving under their parents' expert care the two chicks seemed to know no hardships, developing in leaps and bounds. Within ten days they had changed from mouse-sized, scantily clad white fluff-balls with half closed eyes, barely able to hold up their wobbly heads on feeble necks, to robust chicks in dense, grey down suits. Though still huddled together, they frequently bickered and prodded each other with mini-raptor bills. At four weeks they were already growing pinfeathers and had changed into gangly, leggy young birds. Another week and the two youngsters were scurrying around on the ground, flapping rubbery wings and exercising talons by pouncing on imaginary foes in the leaf-litter. Eager to learn about their expanding realm they were also spending time between meals exploring their surroundings and beginning to look towards horizons beyond the immediate confines of the nest.

In short order we knew they would follow their parents, be shown the tricks of the falcon's trade by having their prey delivered on the wing in air drops to be caught in free-fall, or passed talon to talon somewhere above the forest. As young falcons they would practise mock attacks on branches and fluttering leaves, or dive-bomb each other, flipping upside-down in feigned battle. And with their new-found skills, and a little luck, they should be independent by the end of summer; fully fledged, sleek-feathered, capable hunters of the forest.

Just as falcons have outlived the larger daytime raptors of New Zealand through changing times, so the small morepork (ruru) — a tiny forest owl almost half the size of the falcon — has fared much better than the now extinct, enigmatic laughing owl, a vocal bird with slender legs, standing about 40 cm (16 in) tall, or similar in size to the worldwide short-eared owl. Apparently it liked to roost in caves, where its middens have taught us much about the birdlife with which it shared its favoured semi-open habitat, including open beech forest and the dry mosaic forests that once dominated the eastern slopes of the South Island. With the last known specimen found dead in 1914, the South Island subspecies of laughing owl outlasted its northern relative until surprisingly recent times. By contrast, the morepork continues to be abundant, particularly in the lush, rain-drenched lowland forests dominated by massive podocarp trees.

There are 17 species of New Zealand podocarp, a primitive line of predominantly

Southern Hemisphere pines whose 'cones' do not develop in the conventional way. Instead, they produce single seeds often attached to bright, fleshy, resin-scented fruitlike structures that are irresistible to New Zealand pigeons and other birds. All but two or three grow into stately trees, the majority of which tower above the richly tangled West Coast forests laden in heavy garlands of epiphytes. Like dense palm fronds, kiekie grows here in great aerial thickets swaddling entire trees, while the twisted cables of supplejack vines span the void between ground and canopy resembling tangled ship's rigging from which dangle enticing clusters of bright red berries much soughtafter by many forest birds.

Once common and widespread, the red-crowned parakeet, kakariki, is vanishing from mainland forests but it is still common on offshore islands.

The tallest of all podocarps is the kahikatea whose rigidly straight trunks and fluted grey-green crowns stretch up 60 m (200 ft) or more. They are water-loving trees often at the vanguard of the sodden Westland forests marching into shallow lakes left by retreating glaciers. Cabbage trees and flax help them rebuild the land, followed closely by more complex and mature conifer/broadleaf mixed forests where robust rimu dominate. These are only about two-thirds as tall as the kahikatea but with solid, untapering trunks and wispy trailing foliage. Ferns abound in this realm too. There are tiny ferns and giant ferns, climbing ferns and dangling ferns, translucent filmy ferns and prickly shield ferns. None are more spectacular than the largest of all the tree ferns, the mamaku, growing to 20 m (65 ft) high.

Small insectivorous birds work these forests in busy mixed-species flocks, much as their cousins do in the distant Amazon jungle. Two unique little birds are completely confined to the South Island, the yellowhead, or mohua, and the brown creeper, or pipipi. Both like to congregate, especially when not nesting, to feed with warblers and fantails, busily inspecting foliage, moss, bark and ferns in their diligent search for any and all small insects. They seem to share the food supply in subtle ways that minimise competition. Yellowheads are larger and focus their foraging efforts mainly in the upper canopy, especially in forest types where beech is prevalent. Brown creepers, on the other hand, are inveterate generalists whose constant fossicking takes place at every level, and at various altitudes, blending nervously into the deepest forest shadows. Only the most astute of insects can survive the passage of a flock of these hyperactive, ravenous little predators.

In spite of the cool South Island climate, these temperate forests, whether low and lush or higher and drier in the case of beech forests, remain perpetually green, ensuring a resident birdlife that need not migrate with the seasons. A whole range of birds apportion these forests between themselves in various ways, although their cheeps and twitters, whistles, chatters, squawks and cackles have been silenced in areas where alien predators are taking advantage of their defenceless lifestyles and slow breeding programmes.

The recently recognised orange-fronted parakeet is one such species teetering dangerously close to the brink. Once fairly widespread across the South Island, it now numbers only 100 to 200 birds, mostly retrenched to the chilly central beech forests of Arthur's Pass National Park and Lake Sumner.

With a long history of owls and raptors as their only threats, a high proportion of native birds — parakeets included — nest in tree cavities or on the ground, poor defences against scrambling rats and slinking stoats. The trees, too, suffer from the relentless onslaught of introduced mammal browsers against which they have no defences. Sika deer from Asia, red deer from Europe and white-tailed deer from North America, along with pigs and goats and other escapees from farms, stress the understory. And under the cover of darkness some 70 million brush-tailed possums, marsupials introduced from Australia for their plush fur, attack the forests at all levels, consuming an estimated 21,000 tons of foliage per night. Worse still, they join rats and other predators in depleting bird nests and devouring helpless native forest invertebrates, such as the giant *Powelliphanta* snails. A single possum can dispatch as many as 60 of the slow-growing giants in one or two nights, a recently developed behaviour.

Travelling south across Foveaux Strait to Stewart Island brings relief from this alien oppression, as many of the introduced mammals present on the South Island are still absent here. Suddenly, red-crowned and yellow-crowned parakeets, or kakariki, appear in flocks as they once did across all of New Zealand, and New Zealand pigeons redouble in abundance, calmly plucking buds and berries at eye level. Kaka, New Zealand's large

LEFT Named after its unusual call, the tiny morepork is an adept hunter, preying on insects and birds alike.

RIGHT Acrobats of the forest, kaka use their massive beaks to climb and pluck berries, as well as tear into rotting logs for grubs. They also lap nectar with special brushy tongues, serving an important role as pollinators.

forest parrot with its burnished oxide hues of dark green and red, frequent the local gardens. Both species of New Zealand bat — long-tailed and short-tailed — share the island with all three species of New Zealand penguin, who come ashore to nest in the forest alongside kiwi.

A stone's throw from the main shore on tiny Ulva Island, tucked in Stewart Island's Paterson Inlet, normally arboreal birds boldly drop down to feed on the forest floor. Thanks to the recent eradication of rats here, the leaf litter once again abounds with the seeds and other foods they covet. Kaka even fly across the water to spend time carving up rotten logs with their formidable, oversize beaks in search of fat huhu grubs — wood-boring longhorn beetle larvae. Here, too, the severely endangered South Island saddleback is also making a comeback after being reintroduced from a remnant population on Big South Cape Island.

The sodden forests of the South Island, where rainfall may exceed 10 m (30 ft) in places such as Westland and Fiordland, feels like a world unto itself. Wherever I wander deep into the West Coast wilderness, it is easy to imagine myself stepping far, far back in time. Waterlogged cushions of moss and liverwort squelch underfoot as I walk in the midst of plants that once fed dinosaurs and continue to shelter creatures whose history is too dim to grasp. In the heavy swoosh of a pigeon's wingbeats, the manic squeals and yodels of socialising kaka, the scream of kiwi and the morepork's murmur, I can hear the sounds of the ancient past with me today.

RIGHT The forest-clad ramparts of Fiordland National Park experience few extremes in temperatures, but the high annual rainfall, starting at 6.5 m (22 ft) at sea level and increasing with altitude, creates mist-shrouded scenes little changed through the millennia.

BELOW LEFT AND RIGHT Deeply carved by the last ice age, precipitous peaks preside over the forest-fringed walls of Milford Sound.

LEFT Extremely slow colonisers, beech forests, such as this one lining the steep Haast Valley, take lengthy periods to reclaim land that was once glaciated, unlike broadleaf forests lower down.

RIGHT The diminutive rifleman is the only member of the New Zealand wrens that is still widespread in the beech forests of the South Island.

BELOW Once found throughout the South Island in flocks of up to 200, the yellowhead is declining fast due to predators. Nesting in tree cavities, young remain with their parents in family groups until the following spring, searching for insects under moss and bark.

In the heart of Kahurangi National Park, a female New Zealand falcon incubates her eggs on the forest floor while her mate stands watch. Efficient hunters, they are able to take prey heavier than themselves, such as native pigeons. The pair stays together year-round, courting in mid-air acrobatics and defending a large hunting territory.

FOLLOWING PAGES (LEFT) The mixed podocarp and beech forests of Westland are prime falcon habitat. (RIGHT) Falcon family life is a mixture of tender care and violent death. Parents share duties; the small male delivering prey to his larger mate, which she serves to the chicks in small morsels. As they grow, the chicks learn to kill nestlings offered live and soon begin to investigate the forest floor.

LEFT Besides buds and shoots, broadleaf forests supply innumerable berries to sustain many New Zealand birds, including the large New Zealand pigeon.
ABOVE Bellbirds take insects and berries, but have a penchant for sweet foods, such as honeydew from aphids infesting the trunks of black beech in the northern South Island.
BELOW Even small flowers such as those of the lowland five-finger produce enough nectar to be part of the bellbird's diet.

ABOVE Detail of weeping rimu foliage.
BELOW A kaka uses its formidable beak to browse in the canopy.

ABOVE A kahikatea bearing its typical resin-rich fruit, which is popular with forest birds.
RIGHT Giant podocarps line the shores of Lake Moeraki in Westland, heavily laden with nectar-bearing rata vines.

BELOW A rare South Island saddleback on predator-free Ulva Island works the scaly trunk of a tree fern for insects.

ABOVE Before predators were introduced, the tiered layers of rain-drenched West Coast forests resonated with mixed feeding flocks of small birds, which cooperated in their foraging activities.
RIGHT Sitting motionless, a male tomtit watches for insect movement.
CENTRE RIGHT In one study, red-crowned parakeets were recorded taking 57 distinct food species.
FAR RIGHT A natural predator of small birds and insects, an endemic morepork nests deep within a hollow tree.

RIGHT Straight trunks and small crowns are the hallmark of coastal rimu forests in south Westland. The largest rimu may live 1,000 years and reach 50 m (160 ft), but are more commonly around 35 m (115 ft). Its tiny, primitive cones are rich in nutrients upon which the endangered kakapo depends for breeding success.
BELOW A tomtit hunts amid drooping rimu foliage.

LEFT Brown creepers travel in large family flocks at all forest levels, including dense *Dracophyllum* understory.

ABOVE LEFT Swamp-loving kahikatea trees advance across many Westland lakes, reclaiming the scars left by retreating glaciers.
ABOVE RIGHT Waterfalls slice through near-vertical forest down the precipitous walls of Milford Sound in Fiordland, a testament to the sculpting power of glacial ice.
LEFT The dark, ancient forests of the south draw their lineage from the Jurassic period, having changed little since that time. Of the 17 New Zealand podocarp species, rimu stands out with its drooping foliage, while the conical crowns of kahikatea tower over stockier yet still massive totara and matai.

ABOVE LEFT Stewart Island, still free of introduced stoats and possums, remains the kaka's stronghold.

ABOVE RIGHT In typical parrot fashion, a kaka clambers through the treetops using both beak and feet. In spite of its size, it suffers from competition by wasps and possums for sweet foods like honeydew and nectar, and also falls victim to alien predators.

RIGHT Its dark coppery plumage blending with the forest foliage, a kaka reveals brighter colours when displaying to others in a social group.

ABOVE Described by early pioneers as swooping down on crops in vast flocks, the attractive red-crowned parakeet is now common only on Stewart Island and smaller pest-free offshore islands.

LEFT A red-crowned parakeet seeks tiny ground-hugging bidibidi seeds on subantarctic Enderby Island, where rats are absent. The highly endangered orange-fronted parakeet, newly discovered in localised tracts of South Island beech forest, has similar habits.
RIGHT The giant mamaku, the tallest of New Zealand's ten species of tree ferns, dwarfs the much smaller wheki tree fern, creating tiers of forest habitat in the Oparara Basin.
BELOW Slender gully ferns, growing to 20 m (65 ft), emerge from a gap in the forest of the South Island's upper West Coast.

4

VOLCANOES AND WATTLEBIRDS

DYNAMIC NORTH ISLAND

An opaque, mushroom-shaped cloud rose quickly above the horizon, followed within minutes by a second, then a third. The only blemishes in the otherwise luminous sunrise over the glassy sea of Golden Bay, I realised their significance with a shock: somewhere a volcano was erupting — but where?

A compass bearing transposed to the map of New Zealand gave me the answer: Mount Ruapehu had just awakened from its restless slumber in the centre of the North Island. Exactly 290 km (180 miles) away, the rapidly rising clouds of volcanic ash were visible in every detail from my bedroom window on the shores of the South Island.

ABOVE An icon of shrunken northern forests, a North Island kokako peers from dense foliage, its future dependent on control of alien predators.

LEFT Active for the past 260,000 years, 2,797-m (9,176-ft) Mount Ruapehu and an old rimu tree together denote a vivid sense of time: the land-building forces of our restless planet contrast with the enduring character of the slow-growing podocarp, taking centuries to mature.

Mark and I quickly piled camping and camera gear into the car and headed for the inter-island ferry, arriving at the foot of Ruapehu late that night.

We spent the next five days, and nights, in a land of fire and brimstone, witnessing our planet being rebuilt in front of our eyes. Paroxysms of activity seemed to pulsate almost rhythmically; dense, roiling black ash alternating with columns of billowing steam. Flying over the volcano in a small plane at sunset, we could see that Ruapehu's crater lake was bubbling and boiling, while great curtains of heavy particles drifted back to earth with the wind, soiling the mountain's winter robe of fresh snow. At night, thunder and lightning — electrical storms generated by heat and atmospheric friction from airborne mineral particles — ripped through the sky, the loud cracks mingling with the mountain's own deep throaty rumbles and coughs. With each new spasm, lava bombs the size of cars rose high in apparent slow motion, drawing their own incandescent star trails before succumbing to gravity and arching back down to earth. Our sleeping bags laid out in the snow, we slept little, able to tear ourselves away from

this majestic spectacle only once the volcano began to tire as it slipped gradually into another restless sleep.

Geologically, the North Island is very different from the South Island, and, to a lesser extent, biologically and climatically as well. In fact the two islands have been following their own distinct geologic pathways, riding on separate tectonic plates through much of their long history. It is only in relatively recent times that the two landmasses have converged, scraping somewhat uneasily into one another, a rapprochement punctuated by frequent and sometimes violent earthquakes. The South Island is composed of intricately layered and compressed rocks, ranging from granite to schist and limestone, interspersed with coal seams and gold deposits. Its remnant volcanoes are ancient and extinct, and its mountain-building forces stem from collision, faulting and upthrust, none more spectacular than the Southern Alps.

The North Island, on the other hand, hosts a rich field of dynamic volcanoes, many of which are in their youthful prime. In the west there is the perfect cone of Mount Taranaki, a lone sentinel burnished by wind in summer and encased in rime ice each winter. Around Auckland, to the north, dark basaltic lava flows emanate from young cones such as Rangitoto, which erupted without fuss some six or seven centuries ago, creating its island namesake. Of course the central volcanic plateau commands most of the attention. Here Ruapehu presides over an artist's palette of thermal pools, hissing steam vents, fountaining geysers, bubbling mud pots and crystal-spangled fumaroles. Where gasses and rare minerals percolate to the surface in these multi-hued cauldrons, sulphur, lithium, arsenic, graphite and many other basic elements paint and repaint the landscape. Warm geothermal microclimates thrive, nurturing colourful beds of mosses and algae, while lethal puddles lure innocent birds, such as swallows gathering nest-building mud, to their untimely end.

Volcanoes often destroy all within their reach to better rebuild a new earth. Mount Ruapehu has done so repeatedly in recent years, as have its neighbours Ngauruhoe, Tarawera and Waimangu. White Island, set 47 km (30 miles) offshore to the north in the Bay of Plenty, has been erupting almost continually for decades. Yet these modern eruptions are meek and mild compared to the cataclysmic events of the past. The 1996 eruption we watched up close sprinkled less than an inch of fresh ash

High above the forest a churning cloud of ash rises from Mount Ruapehu during an explosive peak of activity.

down-wind of Ruapehu and caused the brief closure of Auckland International Airport. Yet in gullys and road cuttings around the central North Island exposed layers of soil attest to past ashfalls 3 m (10 ft) or more thick, which would have annihilated almost everything in their range.

In 1886, during a five-hour flash of activity, the eruption of Mount Tarawera tore a 16-km (10-mile) rent in the earth's surface, splitting the mountain in half and opening up Lake Rotomahana. But even this pales compared to Taupo's gigantic explosions over the last 30,000 years. This gaping caldera, better known as Lake Taupo, blew up so violently 1,800 years ago that its atmospheric dust affected the climate on a world scale. Huge areas of the North Island were buried in volcanic debris which undoubtedly redrew the history of the ancient forests that grew there. In the far north of the country timber is being 'mined' from the swampy ground, where massive kauri tree trunks have been perfectly preserved for thousands of years, entombed under thick, waterlogged blankets of volcanic ash. Milled as 'swamp kauri', the timber looks as fresh as if the trees have just been felled, even though they may have spent the last 50,000 years below ground.

The dimpled bark of a giant kauri tree sheds constantly to rid itself of strangling vines, such as this northern rata, which will itself grow into a massive forest giant.

The kauri must rank as New Zealand's most astounding tree. It belongs to the ancient araucaria family and, like the podocarps, traces its lineage back some 150 million years to the time of the dinosaurs, before flowering plants emerged to conquer the world. With relatives in the southwestern Pacific and also in South America, the kauri is the only New Zealand representative of this group. Although no taller than the slender kahikatea, the largest kauri can attain massive proportions. Those that outlived the rampant logging of the last two centuries have been deeply revered by Maori culture — and more recently by modern New Zealanders — and given individual names to reflect their unique stature as living patriarchs of the forests. For example, the two Northland giants Tane Mahuta and Te Matua Ngahere of the Waipoua Forest, with trunk volumes calculated at 245 and 208 m^3 (8,652 and 7,345 cu. ft) and respective girths of 13.8 and 16.4 m (45.3 and 53.8 ft), are the largest survivors standing today. Tane Mahuta's estimated age is between 1,200 and 1,500 years and its crown rises to 52 m (170 ft). Yet it would have been dwarfed by Kairaru of Tutamoe Forest between Kaipara and Hokianga Harbours, whose trunk measured 20.12 m (66 ft) around and rose smoothly to 30 m (100 ft) before dividing into its first branches. Kairaru was thought to be over 4,000 years old when man-induced fire claimed its life in 1886, the same year that Tarawera Volcano ripped the earth apart a few hundred miles to the southeast. Another giant, Kopi, was toppled by a cyclone in 1973. Even after this forest lord had fallen to earth it continued to be home to a colony of rare short-tailed bats living in its hollow core.

Now totally protected, only 6 percent of the great kauri forests that once covered 1.2 million ha (3 million acres) of the Coromandel and Northland peninsulas is left. Felling peaked around the beginning of the 20th century, as did the mining of kauri gum — resin balls used in making varnish and other products, as well as a traditional fire starter and chewing gum for Maori — preserved in the ground for tens of thousands of years. Between timber and gum, kauri accounted for nearly 50 percent of the Auckland Provincial export figures in pre-World War I years. The volume of wood that could be extracted from a single kauri tree the size of Tane Mahuta literally boggles the mind: 245 m^3 (8,652 cu. ft) of absolutely flawless timber, compared to the fractional yield of maybe 3 m^3 (100 cu. ft) from an average mature plantation pine.

Kauri, with their rigid stature and laden in epiphytes and primitive bulbous cones, favour leached ridgetop soils, presiding over mixed forests of podocarps including rimu, totara and miro. Smaller broadleaf tree types establish dense canopy layers that are lower yet. In some ways even more bizarre than these famed New Zealand trees are the *Phyllocladus*, or celery pines. With only five species worldwide, three growing only in New Zealand (the others are found in Tasmania, Borneo and the Philippines), they are modest trees harking back even farther in time than the proud giants in whose shadows they grow. Celery pines develop no real leaves, but instead grow flattened branchlets that are part of the stems. These structures, called phylloclades and looking passably like leaves, are arranged symmetrically like normal foliage and shelter clusters of finger-like, primitive cones shedding their pollen to the wind.

In higher regions of the North Island, beeches also have their place, as do mountain cabbage trees and other plants that thrive in the misty hill country such as the Urewera and Tararua Mountains, which lacks the great, rain-stopping ramparts of the South Island. Though still cool in climate, the mixed forests of northern New Zealand express genuine tropical lushness with their tangled vines and prolific epiphytes. Thick blankets of moss swaddle their branches, and snow-white native clematis cascade through their foliage each spring. Like dense palm fronds, kiekie grows in great aerial masses that look like giant birds' nests, while many forms of rata vines sprinkle the canopy with nectar-bearing clusters of red, orange, cream or white fuzzy flowers designed to appeal to tui and bellbirds.

Found only in the North Island, a female stitchbird feeds nectar to her fledgling perched on twisted supplejack vines.

Acting out the role of both tree and vine throughout the North Island, the northern rata is one of those many remarkable plants that germinates high on the branches of established old trees, taking advantage of a place in the sun without actually stealing nutrients from them. Growing like a vine at first, the rata then sends tendrils downward that soon reach the ground. These thicken and harden as they become woody, the rata eventually overpowering its host and turning into a massive tree in its own right. Each Christmas, rata groves turn crimson, bedecked in bountiful offerings of nectar.

The gnarly pohutukawa, however, steals the show. Its sprawling, twisted trunks, beard-like aerial roots and masses of scarlet-to-rusty coloured flowers overhang many a rocky northern shore. Like the kauri (and also the stunted mangrove forests of the North Island's tidal flats), its natural habitat stops abruptly in a line at about 38° S. At night, geckoes and bats are drawn to the pohutukawa's rich nectar, as are birds in daytime, while bees transform this delectable offering into the whitest and sweetest of all honey. Australian possums attack not only its flowers but also its foliage, causing major dieback in many areas. Making matters even more precarious for this icon of the northern coasts, its seeds are minuscule, 90 percent infertile, and unable to germinate in grass or other thick ground cover.

The dense subtropical forests of the far north are home to North Island brown kiwi that lurk in near impenetrable undergrowth, as do some of the remnant populations of the two rare frogs, Hochstetter's (from East Cape to lower Northland) and Archey's (now found only on the Coromandel Peninsula and in one site west of Te Kuiti). All of New Zealand's nocturnal skinks of the *Cyclodina* genus are restricted to the North Island, along with the splendid Northland green gecko, a member of the beautifully patterned *Naultinus* genus of diurnal forest geckoes.

The birds that feast on nectar, fruit and insects in the canopy were once widespread here too, but their cheeps, twitters and ethereal melodies have all become increasingly sporadic in recent times. Both the gregarious, mainly insectivorous whitehead and the highly territorial stitchbird used to occur throughout the North Island, but their range has shrunk in response to habitat loss and predator pressure. The stitchbird, in particular, dwindled until it survived only on Little Barrier Island well offshore, in the Hauraki Gulf. But with the eradication of rats and possums from a number of other island reserves, populations of both species have been re-established, expanding their numbers once again.

Little Tiritiri Matangi is one of these island sanctuaries — a sheep paddock-turned-Noah's Ark of rare and recovering New Zealand species. From 1973, the returnees have included red-crowned parakeet, whitehead, stitchbird, fernbird, North Island tomtit, North Island robin, brown teal, little spotted kiwi and tuatara. From 37 stitchbirds released in 1995 almost one hundred pairs now divide the island in hotly contested territories thanks to intensive care, including supplementary feeding stations and control of nest-invading mites. New Zealand pigeons and kaka come and go unassisted, some choosing never to leave. But one newly transferred North Island tomtit astounded everyone when it promptly flew home to its mate 56 km (35 miles) away where it had been captured in the Hunua Ranges two months previously. Open to all visitors, and only a short boat ride from Auckland's busy waterfront, many thousands of people come here to steep themselves in a world of birds something akin to prehuman times. When all but a fortunate few have gone again each evening, the island rings with the piercing night-cry of the diminutive little spotted kiwi joining the soft call of moreporks, cooing petrels and mewing little blue penguins.

At dawn, not long after the tui, bellbirds and robins begin hailing the first hint of a new day with their limpid notes and soulful tunes, new voices chime in — strange, other-worldly voices we came frighteningly close to never hearing again. Strident, assertive saddlebacks begin argumentative vocal duels, challenging the airwaves with staccato 'Yak-yak-yak-yaks' in ever longer and louder volleys. An energetic bird, charcoal black with a bright rusty 'saddle' across its back, appears out of the shadows. It moves jerkily, cocking its head up slightly and frequently flicking its tail as it hops jauntily through the low tangle of vines and branches. Most striking of all, delicate pinkish-red fleshy lobes dangle from the sides of its cheeks. A low-level feeder, it uses its stiletto beak to turn over leaves, peel away bark and prise apart roots, mosses and rotting plant matter in search of insects with as much precision as it probes the cup-shaped pohutukawa nectaries for their tiny pools of sweetness. This returnee comes from Taranga Island in the Hen and Chickens group some 80 km (50 miles) to the north, where the entire species was down to just 500 birds or so. Since 1984, when 23 were released into Tiri's scruffy, barely regenerating forest, they have exploded to outnumber the parent island.

The North Island saddleback is, with the kokako, one of only two surviving wattlebirds, an ancient group whose songs bring the forest to life.

Last to join the morning bird symphony comes the rarest and most stunningly beautiful of them all — the kokako. Hauntingly melodious but also deeply mournful, long notes waft softly through the canopy, a song so gentle and peaceful I feel as though it expresses all of the troubles that have befallen New Zealand's birdlife through the centuries. The pioneers called it the organbird for the soulful sounds with which it filled the cathedral forests.

With sky-blue wattles decorating its black-masked face, the slate-grey kokako is the saddleback's cousin, one of three species of wattlebirds, perhaps the most esoteric and obscure of New Zealand's ancient lineage of large passerines. Both are poor fliers that prefer to bound through the forest on dexterous, long but sturdy legs. Kokako live in lifelong pairs, defending 8-ha (20-acre) territories and feeding mainly on leaves, fruits and flowers from forest floor to canopy.

The kokako's tuneful song, performed in distinctive dialects from region to region, once rang heartfully from the tops of the tallest kauri to the fiery pohutukawa coastal fringes. But by the time protection measures were ratcheted up in the last decade, kokako survived only in straggling numbers scattered in lowland forests of the North Island. The remaining 'pairs' often consisting of two males since most of the females had long since been killed by introduced predators as they tried to protect their nests. But when helped with rat poisoning campaigns and possum control, they started to breed successfully again, their numbers increasing rapidly.

Both saddlebacks and kokako were once widespread, each with a distinctive North Island and South Island subspecies. Of these the South Island saddleback is slowly recovering on predator-free offshore islands, such as Ulva. Such efforts, however, have come too late for the South Island kokako. Sporting striking dull orange wattles instead of blue, only a few sightings have been reported in the last half century, and the majestic podocarp forests of the south seem doomed to silence where its rich song once rang.

The third wattlebird, the spectacular and much-celebrated huia, originally found only in the North Island, was perhaps an even more extraordinary species. Mated for life, straight-billed males and sickle-billed females worked the forest tiers cooperatively using their distinctive bills to complement each other in their hunt for insects. Their beautiful black and white tail feathers were so prized in Maori culture that these were worn as ceremonial adornments only by the highest of chiefs, and stored in special, elaborately carved wooden boxes. Their clear song faded into oblivion early last century, but was remembered vividly by a Maori elder, Henare Hemana, who rendered his best imitation, recorded for posterity in 1954 and now freely heard on the Internet — our last remaining memory of a living species departed forever.

Listening to the kokako's song today is to hear the pure ring of time, to mourn the losses endured, but also to celebrate the magic that is still with us — a magic we can and must preserve.

ABOVE Roiling and rumbling, incandescent blocks of semi-molten rock careen through the air and lightning bolts flash in the electrically-charged ash cloud during Mount Ruapehu's 1996 eruption, while star trails streak the night sky during a six-hour time exposure.

RIGHT An aerial view reveals the volcano's crater lake turning to steam as magma invades from below. Although calm has since returned, recent measurements show increased thermal activity, with the 39°C (102°F) lake poised only feet below the rim of a fragile dam of fragmented rock.

ABOVE A composite andesite volcano, Mount Ruapehu last erupted in 1996, as it has done 40 times in the last 150 years. For several weeks, paroxysms of activity sent billowing clouds of dark volcanic ash some 12 km (7½ miles) high into the atmosphere.

RIGHT Mount Ngauruhoe is one of many slumbering cones that dot the North Island's central volcanic plateau. Its current quiescent state belies over 70 bursts of activity since record-keeping began in 1850.

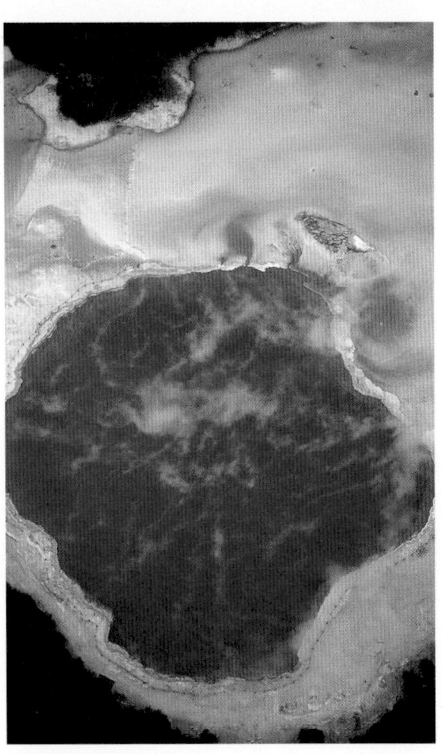

The volcanic workings of the North Island are graphically displayed in the 70 km-long (43½-mile) Taupo Fault Belt, where lakes and pools fizz, hiss and steam as heat, gas and essential elements escape. CLOCKWISE FROM TOP LEFT Among the Waiotapu Hot Springs, the Champagne Pool sparkles with carbon dioxide at a constant 74°C (165°F); a nearby pool is tinted green with leaching arsenic; sulphur bubbles up from a tiny puddle; aerial view of the Champagne Pool; the Inferno Crater in the Waimangu Valley, a geyserlike lake with a pH of 2.1 that rises and falls in a curious 38-day cycle.

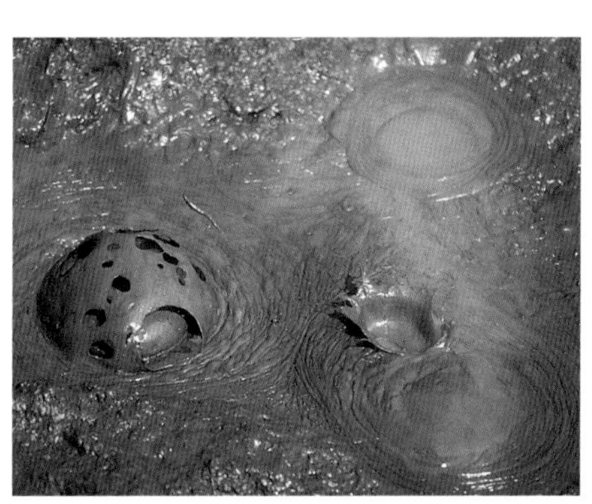

LEFT AND ABOVE Fine graphite splutters away in bubbling mud pots at both Waiotapu and Orakei Korako volcanic areas.
RIGHT The Lady Knox Geyser spews steam and calcium-rich waters for a few minutes each day before settling down again to a contented gurgle.

FOLLOWING PAGES (LEFT) Lightly dusted in winter snow, Mount Ngauruhoe stands asleep while above it airborne ash from its erupting neighbour Mount Ruapehu catches the first rays of the rising sun.
(RIGHT) Like an artist's palette on a geologic scale, volcanic activity tinges the thermal pools of the Rotorua region with hues deriving from gold, silver, mercury, sulphur, arsenic, thalium, antimony, petroleum and even streaks of heat-loving algae.

Surviving only in pest-free environments, such as on Tiritiri Matangi Island Sanctuary, a hubbub of birdlife leads a busy existence among the lacy, silver-backed fronds of tree ferns.

TOP LEFT A national icon, the silver fern, or ponga, spreads its rosette crown deep within the northern forests.

TOP RIGHT Rarely emerging from the thickets, a pair of kokako display with song and dance.

ABOVE LEFT AND RIGHT Once the commonest bird of the North Island, busy flocks of twittering whiteheads systematically uncover tiny hidden insects.

LEFT A lifelong pair of saddlebacks engage in courtship feeding.

FAR LEFT A North Island robin preens beneath shady vegetation.

No flower is too small for the third member of the New Zealand honeyeaters, the rare stitchbird. Sensitive to disease and predation, its brush with extinction has been reversed through intensive care. On Tiritiri Matangi every bird is individually marked and monitored, with only one remaining self-sustaining population on Little Barrier Island. The strikingly coloured male defends his territory fiercely against intruders, while his mate feeds a fledgling chick in the dim heart of the forest. Unusual habits of stitchbirds include occasionally mating face to face and males pairing with several females if nest cavities are close together.

The saddleback (right) and the kokako (above) are both members of the ancient wattlebirds, the oldest lineage of New Zealand birds. They fly little but scramble energetically from forest floor to canopy. Both have distinctive forms in the North Island where they are seriously threatened by predators, though they are not as scarce as their southern cousins, with the South Island kokako sighted only a handful of times in the last 50 years. Wattlebirds mate for life and are splendid vocalists, the soulful kokako is as melodious as the saddleback is assertive. The extinct huia, with sickle-billed female and straight-billed male, was the only other wattlebird species known.

With long legs and dextrous feet, the kokako feeds on leaves, berries, insects and flowers, such as the beautiful wild clematis cascading from the forest canopy, one of nine native *Clematis* species. The kokako's haunting song, likened to organ pipes, varies from place to place in distinct dialects. Sedentary pairs hold territories from 4 to 20 ha (10–50 acres), spending an entire year caring for their young, whose wattles change from pink to blue during this time. Thanks to predator control, the species was returned from the brink of extinction to over 600 pairs and rising.

FAR RIGHT A North Island icon, the splendid pohutukawa tree festoons many a rocky shoreline in the north, such as here at Tiritiri Matangi Island in the Hauraki Gulf.

RIGHT AND BELOW (LEFT AND RIGHT) The saddleback is an active insect hunter, using its stiletto beak to peel away bark, unfurl curled leaves and turn over dead plant matter as it works all levels of the forest, often attracting other birds in its wake. It also relishes the copious nectar available when pohutukawa come into summer bloom, smearing pollen grains on its beak and face. Last seen on mainland North Island in 1910, the species survived on just one island in the Hen and Chickens group, but was reintroduced to eight other islands where it has since thrived. More recently, it was successfully returned to protected mainland reserves, with a total population estimated at 5,000.

LEFT Embattled survivors from another age, enormous kauri trees were once the contemporaries of dinosaurs. Living for perhaps 4,000 years and measuring 20 m (65 ft) in girth, the largest trees were milled in colonial days, when abundant wattlebirds still nested in their lofty crowns. Native birds are now being brought back to the magnificent Cascade Park in the Waitakere Ranges near Auckland thanks to intense predator control.

RIGHT In a typically assertive posture, a saddleback flicks its tail and lowers its head before hopping to the next patch of bush.

BELOW The kokako is a large bird at 38 cm (15 in) long and weighing up to 270 g (9½ oz). It uses its crushing beak to process its eclectic diet.

ABOVE Lake Waikaremoana, cradled at the heart of the rugged Urewera Ranges, surrounded by majestic beech forests.
BELOW Known as kukupa in the north, the New Zealand pigeon revels amid tangled supplejack which provide it with some of the berries common in the mid-canopy.

RIGHT The forests of the North Island exhibit a distinctly subtropical luxurience where vines and epiphytes abound. High in misty Te Urewera National Park, mountain cabbage trees raise their broad-leaved crowns between tall, moss-laden beech trees.

5

FROSTY WORLD OF THE MOUNTAIN PARROT

THE SOUTHERN ALPS

All through the long night, scrapes, screeches and rattles, heavy footsteps and manic chortles have resonated from the tin roof of the alpine hut just above my bunk, making for fitful sleep. At first light, I rub the tiredness from my eyes and, clearing heavy condensation from the windowpane, find myself nose-to-beak with a quizzical, bedraggled parrot face. Laughing involuntarily, my frustration instantly turns to admiration for New Zealand's cheeky kea, or mountain parrot. Since the first pioneers set foot in Aotearoa the kea has been much maligned for its bold and very often destructive curiosity. From the start, its cunning intelligence placed it at odds with the high country farmers who invaded its realm with sheep and fire. Yet one can but marvel at the bird that has left the confines of the forest to make a living in the stormy heights of the majestic Southern Alps. Whether it made this move as a response to the advancing ice age glaciers that spread toward the lowlands and pushed back the forests or to evade the talons of now-extinct forest eagles and harriers will never be known, but that the decision served it well cannot be doubted. In fact, this eccentric choice set it well apart from all other parrots in the world. It is also clear that its exploratory nature and inveterate arrogance are key to its existence in a frosty environment where only the cleverest survive.

Little escapes the kea's attention, and its querying beak is immediately put to work investigating anything that may seem out of the ordinary. The formidable, recurved dagger probes and cuts, lifts and pries anything at all that is either new or remotely edible. Kea naturally dig up roots, tubers and bulbs, turn over stones in search of hibernating insects, even raid other ground-nesting birds. They also peel bark off shrubs and seek out tiny ground berries among the tussock grasses. Above all, they are drawn to check out anything that simply looks intriguing, making short work of tents, bicycles or car window seals. Parked motorcycles are especially enticing with their stuffed seats and

ABOVE A busy group of kea fossick among the frost-flaked schist above Fox Glacier.

LEFT Fox Glacier wends its way from the Southern Alps toward the Tasman Sea.

complex wiring — temptingly easy to unstuff and unravel — and thus irresistible to hyperactive, snipping beaks. The innate object of their search may be edible titbits, but the apparently wanton destruction can be infuriating.

Even more problematic has been the kea's need for fatty foods to make up for the rigorous mountain climate, and its ability to satisfy this need by attacking live sheep. Some individual birds have learned to kill by eating into the backs of snow-bound sheep to reach the fat around their kidneys, a grisly act sometimes carried out on dark winter nights. This trait led to a colonial government bounty being placed on the kea's beak, which, after a century and a half of persecution that cost the species some 150,000 kea lives, was only lifted in 1970. The number of kea surviving today may be one hundredth that figure, with estimates varying between 1,000 and 5,000. Fortunately — or unfortunately depending on one's point of view — kea make themselves naturally conspicuous in many parts of the magnificent snowy ranges of the South Island, and in so doing give the impression of being quite common.

My favourite of all kea locations are the tussock ridges and alpine gardens surrounding Chancellor Hut which overlooks Fox Glacier. Even if I've lost more than one night's sleep listening to their antics on the tin roof. Step outside and breathtaking, sweeping vistas encompass the immaculate peaks of the great divide, the crevassed névé and creaking, fractured glacier far below. Encased between dark rock walls, it tumbles sinuously down toward the lush forest. The Tasman Sea is just a stone's thrown beyond, where the sun sets in golden pools even as late snow flurries descend from the peaks.

Conversely one can walk through luscious groves of kahikatea and southern rata just below the glacier's snout and look back not only at the luminous river of ice filling the valley, but also at the highest ice cornices of the Southern Alps dividing the watershed east and west — stupendous peaks such as Tasman, Cook and other equally breathtaking icons overlooking the wild Westland forests. But even as we watch, the glaciers are retreating almost palpably. Both the Fox and neighbouring Franz Josef Glaciers have shrunk by several miles in the short time since people first started flocking here to admire their grandeur. In their wake a new forest has sprung up harbouring a vanguard of pioneering plants and animals, their adaptability hewn by past glaciation cycles.

Each spring, alpine bogs turn into miniature gardens where tiny carnivorous sundew plants thrive.

Kea are at home looking down on this grand landscape from on high, flying in flocks above icy backdrops, and yodelling across the valley to keep in touch with their mates as they work the herbfields and scree slopes in search of berries and bulbs. Photographing them in this magnificent setting has been an obsession of mine, their playfulness often leading me in frustrating, fruitless circles. No sooner would I follow them into a difficult gully where they were busy shredding juicy herbs, than the whole mob would fly back to where I'd left my camera bag, ripping into it with gusto as I painstakingly slogged my way back up. Or I'd spend days very nearly ruining my gear in rain and sleet trying to capture their playful antics tussling in the wet snow, only for them to depart across the valley the moment the sun reappeared. And when I lay on my stomach for close-ups of their stone-turning sprees on the shingle slopes, they saw it as an invitation to hop onto the backs of my pant-legs and boots, evading my lens every time.

Christmas time around Chancellor Hut also brings explosions of splendid flowers to the kea's realm. As the winter snows recede, several types of hardy mountain daisies of the *Celmisia* group, with their leathery or furry leaves, come into bloom, as do multiple varieties of alpine buttercups. Many other ground-hugging herbs, such as carrot plants and shy edelweiss, spread their delicate petals from hardy roots among the sun-warmed schist slabs. The most eye-catching of them all is popularly known as the Mount Cook lily — in actual fact not a lily at all, but a giant buttercup. Its thick parasol leaves and graceful, immaculate flowers defy the elements with their ephemeral beauty. With 13 to 20 cm (5–8 in) leaves and 1.5-m (5-ft) high stalks, it is the largest member of the Ranunculaceae family in the world.

While Mount Cook lilies favour wet areas with plenty of snowmelt (they are indeed abundant around the base of Mount Cook), other alpine flowers thrive in much drier conditions. Further north, in the mountains of mid Canterbury, Nelson and Marlborough, grow the bizarre plants known as vegetable sheep, along with other cushion plants. Forming rock-hard, velvety masses that hug the ground like grey-green pillows many feet across, they dot the high country up to 2,750 m (9,000 ft) like so many inanimate sheep grazing the sparse hillsides. Other plants prefer loose scree slopes, such as the odd-looking penwiper plant and black daisy. The speargrasses of the *Aciphylla* genus — also called spaniards — raise their viciously spiked heads up to 4 m (13 ft), with different species apportioning the various alpine habitats. Thus we have the feathery spaniard, golden spaniard and horrid spaniard, as well as giant spaniard and dwarf speargrass, to name but a few.

Summer carpets of alpine gentians spread through the tussock grasslands of the Cobb Valley.

Enormous Kahurangi National Park, 4,500 km^2 (1,740 sq. miles) in area, is the hotspot for alpine flower diversity. With its granite outcrops and quartz ridges, pavement sandstone and marble plateaus, mazelike karst formations and trilobite-rich limestones, it provides a multitude of habitats above treeline range, from open fellfields to vast tussock downs. The slopes of Mount Arthur bristle with groves of large grass trees — *Dracophyllum*, meaning 'dragon-plant' — straight out of *Lord of the Rings*, while glacially carved, U-shaped Cobb Valley is a haven for some of the finest alpine meadows anywhere. Each December fields of Maori onion, *Bulbinella hookeri*, bedeck the damp valley floor, dotted with rich clumps of pale gentians.

Some of the most incongruous inhabitants of these high realms are also the least conspicuous. One would not expect to find too many cold-blooded animals living in such a frigid environment, so it comes as a real surprise that the rare black-eyed gecko is an alpine species found between 1,300 and 2,200 m (4,300 and 7,200 ft) high in the northern ranges of the South Island, where its habitat remains under snow for many months of the year. Coloured in the same tones of grey as the rock bluffs it favours, it is mainly nocturnal and can remain active in air temperatures as low as 6°C (43°F). Another alpine reptile found only in the South Island is the scree skink, equally cryptic in colour and living at altitudes up to 1,400 m (4,600 ft). Much farther south, the splendidly patterned harlequin gecko is one of the southernmost geckoes in the world, living in the windswept subalpine scrub and granite outcrops of southern Stewart Island.

Even in these harsh environments many of these little-known, timid species are suffering from attacks by rats or stoats, and probably also habitat loss from introduced grazers destabilising the ground and degrading the alpine vegetation. Even as their existence is becoming more precarious, so new species are being discovered. In 1996 a new, extremely rare gecko was spotted living on scree slopes above 1,000 m (3,300 ft) in the cold wind-blasted Takitimu Mountains of Southland. Even more astounding, geckoes are being discovered in some of the highest ranges of the Southern Alps, even in the Mount Cook area, on rock faces above 1,500 m (5,000 ft). A specially trained border terrier named 'Putiputi rapua' is being put to work by the Department of Conservation to locate as yet undiscovered rare and isolated populations.

Many invertebrates of these mountainous locales are equally fascinating. There are scree weta and scree grasshoppers, tussock butterflies, tussock weevils, tussock cicadas and tussock grasshoppers. In fact 12 of the 15 species of short-horned grasshoppers in New Zealand are found in alpine habitats. New species of weta are still being discovered high in the mountains, such as the bluff weta of the Kaikoura Ranges, with its fiery coloured legs and blue-grey body. Amazingly, in this harsh environment, it is not unusual for these soft-bodied insects to freeze solid during winter.

Also living well above the treeline, between 900 and 2,500 m (3,000 and 8,200 ft), is one of the South Island's most unusual, if innocuous, little birds, the rock wren. At 10 cm (4 in) and around 16–20 g (just over an ounce), it is slightly larger than its minute cousin of lower elevations, the rifleman. Living amongst boulder fields, where it scampers about in search of insects, its lifestyle is uncannily shrew-like.

Rock wrens are quite sedentary, flying only for the shortest distances. They remain in the high country through the winter, happily living in tunnels and airspaces beneath thick blankets of snow in the same manner as do voles in the Arctic regions. The invasion of mammalian predators and competitors has taxed this extraordinary little bird's survival, but luckily, the harshness of its habitat still gives it a slight edge, though its numbers are waning.

The story of another avian oddity has come to us with a different, much happier twist. For over a century the takahe, a hulking 3 kg (6½ lb) pedestrian bird with deep blue and green feathers and a bright red, cleaver-like beak, was known only from four museum specimens. But in 1948, the persistence of a young medical student, Geoffrey Orbell, who doggedly pursued rumours of the existence of a huge flightless swamphen in the remote Fiordland ranges, led him to rediscover this amazing bird high in the rugged, rain-drenched Murchison Mountains. In the nick of time, with a population of just two or three hundred and dwindling, protection measures were trialled and tested, and captive breeding programmes started. Introduced red deer that stripped its food sources were culled and predatory stoats controlled. Small breeding populations of takahe were also established on several offshore island sanctuaries that were free of these threats.

The natural habitat of the takahe as we know it today consists of high-altitude tussock grassfields and beech-lined hanging valleys, where the birds feed mainly by pulling out snow tussock leaf bases and digging up fern rhizomes. But subfossil bones and Maori midden remains show us that they were once widespread throughout the lowlands. It was no doubt the arrival of humans in New Zealand that caused them to retreat behind their mountainous ramparts. Even in the comparatively balmy climate of Tiritiri Matangi Island, the takahe has adapted well to coastal conditions once again, favouring bracken areas and feeding heartily on grassheads and stems. Pulling out the tasty bits with its massive beak, it then grabs the stems between its sturdy toes for a firm hold while it strips and chews the edible base.

In every respect the takahe looks and behaves like the sumo wrestler of the bird world, especially when it comes to territorial defence. Males can engage in fierce battles that may recur over several days, and sometimes even attack people in their eagerness to guard their turf from intruders. Living for 20 years or more, they are thought to form lifelong pairs, with grown young often helping their parents raise new chicks. But the dozen or so pairs living on Tiri seem to spend more time jumping borders and swapping mates than anything else, so much so that the closely kept takahe roster on the island reads like a high-drama soap opera. Perhaps this stems from the relatively small territories they hold here, rarely exceeding 4 ha (10 acres), compared to 20–30 ha (50–75 acres) in the alpine tussock habitat of the Fiordland mountains.

ABOVE AND BELOW Two odd birds of the Southern Alps: the hulking takahe and the cheeky kea.

It is amazing to realise that for nearly a century this wonderfully preposterous bird figured in books on cryptozoology (the science of barely known and doubtfully recognised species) before making a comeback with the help of vigorous Department of Conservation management. Indeed, from oversize gallinules to undersize wrens, snow-loving parrots to heat-shunning reptiles, the snowy mountains of the South Island in their still-life grandeur continue to hide many a facet of adaptation and survival that stretch our knowledge of biology as well as our imagination.

ABOVE At 3,036 m (9,961 ft), the knife-edge summit of Mount Aspiring catches the first rays of sunrise.

RIGHT As the sharp contours of the Southern Alps — here framed by towering kahikatea trees — have emerged from the embrace of the last Ice Age, so the magnificent podocarp forest has reclaimed the precipitous valleys, keeping pace with retreating glaciers.

LEFT Icy winds and snow flurries do not worry a kea at home in the high ranges.

ABOVE Evening sunshine highlights the contrast between the lush crown of a mature lowland kahikatea tree and the snowy summits of Mount Tasman (left) and Mount Cook, 3,498 m (11,476 ft) and 3,754 m (12,316 ft) respectively.

LEFT Ice falls tumble down from a craggy névé to form the upper Fox Glacier, as it pushes its way west.

RIGHT The Franz Josef Glacier, along with the Fox, are the only glaciers left whose tongues still extends down into the lowlands. Stretching well out to sea during the Ice Age, the Franz Josef is now just 13 km (8 miles) long and actively receding, having retreated by several miles in the last few decades alone.

ABOVE Rimed with wind-blasted ice cornices, Mount Tasman gives birth to a number of the 3,135 individual glaciers that riddle the Southern Alps, many of them barely a few acres in area. By contrast, the volcanic cones of New Zealand's North Island spawn only 18 diminutive permanent glaciers.

RIGHT As pristine as the snows near which it grows, a giant buttercup attracts a tiny pollinator.

LEFT The Mount Cook lily is in actual fact the largest buttercup in the world, with leaves up to 20 cm (8 in) across and flower heads reaching 1.5 m (5 ft) high. Growing in alpine herbfields where snow melt saturates the tussock, it bursts into bloom during December and January.
BELOW Sun-warmed rock slabs heating their roots speed the development of creeping ourisia along with korikori, one of many types of alpine buttercup.

When winter frosts lose their grip on the unstable fellfields above the tussock grasslands, a profusion of splendid alpine plants explodes from scree slopes and rock fissures just below the snowline, including bristly carrot (right), buttercups (left) and numerous types of *Celmisia* mountain daisy (below).

Persecuted since colonial times and deprived of many of their food sources by fire and introduced grazers, kea have relied on their innate resourcefulness and intelligence to survive in the mountains of the South Island, feeding on berries, buds, roots, insects and even other birds and their eggs.

ABOVE AND RIGHT Some 150,000 keas were hunted over the last 130 years to protect the sheep they sometimes kill. The remaining 1,000 to 5,000 are now protected, as is much of their high alpine habitat.

FOLLOWING PAGES (LEFT) A kea's plumage reveals an array of subtle colours as it flies over a deep alpine gorge. (RIGHT) Kea are rambunctious, social, playful, innately curious and acutely intelligent. Oblivious to harsh weather, they tussle and play in the snow, spending much time socialising and investigating anything that is remotely edible or catches their fancy.

ABOVE The drier northern South Island ranges in Kahurangi National Park have long since lost their ice cover, but the deep U-shaped Cobb Valley, with its tussock tops and beech-clad slopes, attests to past glaciation.

RIGHT All spikes and needles, a spaniard grass, *Aciphylla*, blooms at the edge of a scree slope, a harsh habitat for rare plants and geckoes.

FAR RIGHT Small daisies find shelter under one of the strangest plants of the high ridges of Kahurangi — the 'vegetable sheep', *Raoulia*, so named for its woolly cushionlike growth.

ABOVE With the long days of summer sunshine, vast fields of Maori onion, *Bulbinella hookeri*, paint the valley floors in the mountains of northwest Nelson.

ABOVE LEFT AND RIGHT Cold, swirling mists envelop the rugged mountains of Fiordland. The takahe's last wild stronghold can be found here, deep in the Murchison Mountains.

RIGHT Flightless and built like an avian sumo wrestler — standing 50 cm (20 in) tall and weighing 3 kg (6½ lb) — the takahe was rediscovered nearly 60 years ago after being thought extinct for half a century.

FAR RIGHT Strong feet are used to hold juicy grass stems while mincing them with a sharp beak.

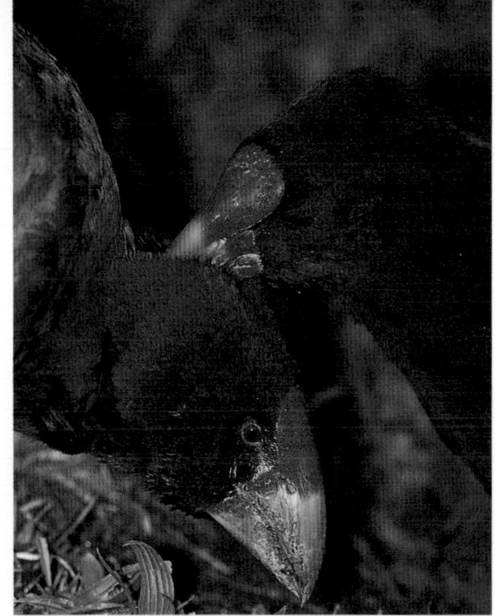

LEFT Everything about the takahe is massive. Its solid proportions stand out as it drinks among the reeds.
ABOVE Cleaver-like beaks are used as effectively in vicious territorial battles as in delicate mutual preening between a pair, seen here.
RIGHT Starch from fern rhizomes and the base of juicy tussock stalks, along with grass seedheads, are takahe staples.
BELOW Parents feed their chicks seeds and morsels of plant matter until they are almost fully grown.

ABOVE Tussock downs on Stewart Island.
LEFT Widespread across New Zealand before the arrival of humans, takahe are pedestrian birds of grasslands, especially dense tussock in the rolling mountains.
RIGHT Frost on red tussock in Gouland Downs, Kahurangi.

6

WONDERS OF THE WATERWAYS

LAKES, RIVERS, SWAMPS AND ESTUARIES

Spring has arrived in the foothills of the Southern Alps, where snowmelt is causing the wide, intricate braided rivers to shimmer in their grey shingle beds. Mount Cook is glowing in sunshine far above, and the Tasman River weaves its multitudinous strands down from the dark, rubble-strewn snout of the glacier of the same name. Just where these silvery ribbons spill from the morainal debris into the cold, glaucous waters of Lake Pukaki, a minor miracle is taking place. Seven graceful black birds with tall, impossibly thin red legs are dancing and zigzagging in the riffles.

I am watching a flock of New Zealand's black stilts, or kaki, one of the rarest waders in the world. With a total of only 84 individuals in the wild, and another 20 held for captive breeding (at the time of writing), it truly feels like a miracle that I should be watching seven of them here today. They are busy birds, constantly on the move, tipping their bodies rhythmically forward as they pluck tiny insects from the clear flowing water with ever-so-slightly upturned, needle-like beaks. Working loosely in pairs, they are also talkative, their high-pitched 'tweeeeks' ringing on the cool breeze as they take off on brief flights, drawing close loops over the shingle, chasing and feinting in what appears to be mild aggression between neighbours. An eighth bird, behaving similarly, is smaller and mottled black and white, probably a hybrid resulting from a miss-match between a black stilt and its much more common and widespread relative, the pied stilt, which sometimes happens when the rare birds can't find a mate of their own kind.

The black stilt's troubles, like those of so many other New Zealand birds, come from invasive predators and, in this case, also introduced weeds like lupin and willow that are altering the ecosystem of the braided rivers. Water diversion for irrigation and hydroelectric power adds to their woes. This is why a captive breeding facility is being run

ABOVE Pied oystercatchers flock to the shell beds of Awaroa Inlet in Abel Tasman National Park.

LEFT Stormy weather descends onto the Matukituki River flowing from the Southern Alps toward Lake Wanaka.

by the Department of Conservation in Twizel near Lake Pukaki, successfully producing several dozen hand-raised stilt chicks each year to add to the wild population.

On this clear, breezy afternoon the stilts are not alone. Blending in among the pebbles, banded dotterels are scurrying about, also picking invisible insects from the rivulets' edges, peeping and bobbing, many of them just returned from winter in Australia. Very soon wrybills will also be arriving into the upper reaches of the braided river system to breed, having spent the winter in estuaries mainly in the North Island. They are the world's only bird with an asymmetrical bill; bent to the right to help it turn over pebbles in search of food.

Common in waterlogged grassland, a pukeko spreads its stumpy wings.

Spring is in the air all over the landscape, and nowhere more so than through the extensive network of wetlands which represent the richest and most diverse wildlife habitat in New Zealand. Farther down the braided river system, black-fronted terns and black-billed gulls are readying themselves for nesting in colonies on stream banks and small islands. They are returning inland after leading a coastal existence in winter, unlike their more maritime relatives the white-fronted terns and red-billed gulls which will remain there. Larger lakes and tranquil ponds, especially in forested lowland areas, are being claimed by grey ducks and grey teal which move more freely outside the nesting season. Although sometimes confused with the more common and far more conspicuous introduced mallards, they prefer calm lakes and slow-flowing waterways. Two other birds of inland lakes are the great crested grebe, widespread around the world but found in New Zealand only in the South Island, and the small endemic dabchick, now restricted to the North Island. The shy New Zealand shoveler, a subspecies of the Australian shoveler, though numerous and highly mobile throughout the country, seeks out even more private environs surrounded in reeds and rushes. Most secretive of all is the brown teal, or pateke, a critically endangered New Zealand endemic with the dubious distinction of being the world's rarest duck.

With its abundant rainfall — 1–2 m (3–6 ft) annually over much of the country, reaching 10 m or more (33 ft) in parts of the Southern Alps — New Zealand is replete with myriad varied and productive wetland environments, the richest of all bird habitats. Lakes abound, starting with the huge glacial lakes left in the wake of the last ice age, such as Hawea, Wakatipu and Te Anau nestled among the Southern Alps. At least 750 lakes measuring over 800 m (half a mile) across are sprinkled all over the country, plus innumerable swamps and wetlands wherever drainage is poor. In addition, literally thousands of miles of streams and rivers drain the mountainous landforms like a filigree network. None of these are huge, but most flow vigorously on their short journeys to the sea. Some waterways take massive detours underground through a honeycomb system of tunnels, caves and deep erosion channels that riddle entire limestone mountains, re-emerging as some of the clearest springs in the world, such as Waikoropupu Springs — Pupu for short — in Golden Bay. These clean-flowing river systems are inhabited by many interesting native fish, none more so than the giant eels. These amazing fish may live for 100 years or more and reach 1.2–1.8 m (4–6 ft) in length and weigh over 24 kg (50 lb). At the end of their lives they leave their freshwater home and travel thousands of miles into the tropical Pacific to spawn.

There is no bird who loves the rushing waters of cold, clear torrents tumbling through wild forested gorges more than the increasingly rare blue duck, or whio. With strong feet and strange fleshy flaps on the sides of its beak, it is a specialist living where any other duck would be swept away. Pairs are territorial and guard their stretch of river jealously. Searching for cadis fly larvae on the eddy side of rocks and boulders, even newly hatched ducklings are adept at scuttling over frothing cataracts with their large paddling feet, or diving beneath swirling currents and standing waves. There are only a few examples of similar adaptations in the world, notably the torrent duck in the Andes of South America. But none show the bizarre lip-like extensions to the edges of the beak which apparently serve as a kind of foil to prevent the fast-flowing water from washing away their food as the prey is lifted from the uneven stream bottom.

But blue ducks, like so many other unique New Zealand birds, have not been able to adapt to predators and to the degradation of their habitat. With forest clearance and increased pasturelands, the clear streams they rely on are more prone to flooding and silting. And, being doggedly sedentary, the shy birds will not leave their territory when either they or their habitat is disturbed beyond repair. Although still widespread, only a few thousand remain, their population boosted by conservation projects in several areas.

Another delightful endemic little duck is the New Zealand scaup, whose glossy black, bathtub-toy shape adorns deeper ponds and lakes throughout the country. They are adept divers and forage busily along the bottom, leaving only a trail of bubbles as a hint of where next they might reappear. The high-pitched, bell-like whistle of the iridescent males often adds a delightful touch to mirror-smooth ponds and lakes in the early morning.

At lower elevations, swamps, estuaries and mudflats offer yet another extension of the varied wetland systems. A number of New Zealand's iconic plants, which figured very significantly in the Maori livelihood of pre-European times, also grow here. Raupo is a bulrush growing in standing water whose every part was used in daily life: fronds for thatching, roots for cooking, pollen for baking into cakes, and, in more recent times, the downy fluff from the seedheads was used by the early pioneers as mattress stuffing.

Widespread across the swamplands, large elegant bunches of toetoe grasses, graceful relatives of the commonly planted South American pampas grass, send up banner-like golden heads waving in the wild spring winds. Maori used to fly intricately shaped kites made from their stalks. Very often these 3-m (10-ft) high grasses grow intermixed with tall, hirsute cabbage trees and dense New Zealand flax. Called harakeke by Maori, the latter belongs to the agave family, much more widely known around the world for its many hardy desert forms. With its bladelike grey-green leaves and profusions of flowering stalks heavily laden with nectar, flax was probably the most influential plant of all in early New Zealand history. Its fibres and leaves were woven into many of the most important Maori daily implements, from footwear and clothing to baskets and fishing nets. Its dry stalks were made into rafts, its copious nectar served as a culinary sweetener, while the plant was also valued for medicinal uses. Superbly strong flax ropes also found a place in the world market, and continued to be in demand up to the 1930s.

Several of New Zealand's least conspicuous birds inhabit these marshes, particularly the spotless crake, which emerges only ever so fleetingly from the security of dense reed beds, while the marsh crake favours salty coastal zones. The rather larger banded rail confines itself mainly to rushes and mangrove flats in the warmer north. The endemic brown teal is a secretive little creature of flax swamps, which it shares with the elusive Australasian bittern, whose presence is often revealed only by its booming call, heard on calm nights over a mile or more away. The New Zealand snipe, too, favoured marshy habitat but, unlike its cosmopolitan relatives in other parts of the world, was quite sedentary and vanished with the arrival of mammals except on outlying subantarctic islands.

A weka emerges from the wet forests on Ulva Island to hunt invertebrates along the shoreline.

Where the rivers flow into the sea there are numerous, extensive, broad and sometimes complex tidal wetlands that attract huge congregations of birds. A large number are migrant waders who flock here after travelling from the far reaches of the Arctic to escape the Northern Hemisphere winters. Ruddy turnstones, lesser knots and many others gather in their many thousands on favourite sandbars and mudflats, very often where fresh and salt waters mix. Others, like the recently arrived black-fronted dotterel, are resident year round. White-faced herons are also commonly seen here, and much more rarely royal spoonbills and white herons, both stunningly beautiful birds in all-white robes. The white heron, revered by the Maori as the kotuku, but ranging throughout the world as the great egret, has but one nesting colony high in the kahikatea forest overlooking the Okarito Lagoon in Westland. The budding resident population of spoonbills joined it here in recent times and has expanded to form other successful colonies.

The most conspicuous bird of the estuaries is without a doubt the South Island pied oystercatcher. Wherever the sand or mud is rich in clams and other invertebrates, these striking birds abound. But they are by no means tied to the tidal systems, and quite happily venture into soggy fields and golf courses, or any waterlogged areas soft enough to thrust their probing, dagger-like beaks deep into the ground in pursuit of earthworms and shellfish alike.

Wet fields and open grasslands are also the premises of several other striking birds. The masked lapwing, or spurwing plover, with its odd-looking bright yellow facial shield, is widespread in Australia, New Guinea and beyond. Like a number of other birds from silvereye to spoonbill, it made its debut in New Zealand in tandem with the changing landscape brought about by human settlement. Expanding pasturelands have provided it with a perfect habitat, where it picks up surface insects with its short bill. Pairs are aggressive and noisy when nesting, and relentlessly pursue any passing Australian harrier, another widespread bird that favours swampy grasslands to hunt and nest.

With its closest relatives in Australia, South Africa and Asia, the beautiful paradise shelduck, or putangitangi, is one of the more spectacular of New Zealand's endemic birds. Sadly, it is often unappreciated simply because it is a conspicuous part of the everyday landscape. Even more tragic, it is one of the few New Zealand endemics which is still hunted in their tens of thousands, and attracts disfavour because it consumes pastures or soils golf courses, harking back to the bad old days when persecuting the native fauna was widely regarded as the 'rightful' thing to do.

In an unusual reversal of roles, the female paradise duck stands out from the male with her pure white head.

Yet paradise ducks have a fascinating life. After hanging around in flocks for their first two or three years of life, the young birds get together to form lifelong pairs who keep each other constant company. Mates are easily distinguishable by their strikingly different plumage, their loud *zeek* and *zonk* calls alternating as they fly close together. The female is a beautiful rufous, with white head and metallic green scapulars and has the lower voice, whereas the more dapper male is finely patterned in chestnut and metallic black hues. The ducklings, proudly guarded by their vigilant parents, are equally attractive; busy fluffballs boldly streaked in black and white down.

Paradise ducks, referred to by hunters as pari for short, can live for over 20 years, but with over a third of the entire adult population dying each year, their average life expectancy is reduced to less than two and a half years. Yet, given the right circumstances, the birds can also form lasting relationships with humans, demonstrating all of the faithful dedication to a particular person as they do between mates. Such was the case of 'Daphne' who lived with her wild mate on Tiritiri Matangi Island, but was at the same time so deeply attached to her human friend that she would sometimes fly alongside the ferry when this person left the island. Daphne inspired national sympathy and anti-hunting fervour in 2004 when she was heartlessly shot as a golf course 'nuisance', together with her moulting flock, while unable to fly.

Happily for wildlife enthusiasts, and in spite of such persecution, paradise duck numbers remain high. Unlike many other species, they benefit from increased habitat through expanded farmland, and thus promise to grace the New Zealand wetlands securely into the future. But the question remains: Why are people prone to have far more appreciation for rare and dwindling species than for those that are common and easy to observe?

The most versatile and successful of the swamp birds is the pukeko, ranging widely through Africa and Asia as the purple swamphen, and closely related to the American purple gallinule. In addition to its worldwide relations, it also bears an obvious kinship with New Zealand's larger, flightless takahe. However, the pukeko distinguishes itself in its highly unusual social behaviour, where up to a dozen related birds band together into a breeding family. Sometimes only the dominant pair lays eggs, and all the other members of the group, generally consisting of offspring from earlier breeding, help out. But in some cases the mating scheme is more equalitarian, including several males and females sharing mating and chick care, with up to 20 eggs laid in a communal nest which is looked after by all.

While the takahe is thought to have descended from an ancient lineage of gallinules, or swamphens, that adapted to the prehistoric New Zealand ecosystem, the pukeko appears to stem from a relatively recent arrival. Living primarily around swamplands, it readily takes advantage of environments created by people's activities, particularly well-grazed farmland, and has thus been able to withstand often wanton and merciless persecution.

Another oft-maligned wetland bird has been rather less fortunate. The weka is a very large, and also very clever, flightless rail that was until recently widespread and common over much of the country. An opportunistic bird, it is able to make a living as comfortably in forests, flax swamps or farmland as along any shoaling coastlines. It acts as a predator of other birds' nests as easily as a scavenger of the tideline detritus and backyard compost piles. With a spirit akin to the kea of the high mountains, the weka will happily steal a cat's bowl or picnic-table teaspoon in the event that these might be edible, or run off with a tea towel for closer investigation in the sanctity of the garden hedge. Some of my fondest first impressions of New Zealand when I moved to Golden Bay 15 years ago were learning to cope with the resident weka family. They very soon recognised a soft spot and would dash into the house whenever the front door was left open, raiding apples from the pantry or bread from the breakfast table. When I planted my first vegetable garden, most mornings I would find a weka following about three paces behind me, pulling out the seedlings that I had just carefully planted. One pair and their three chicks held court on the front lawn, while another family worked the woodpile at the back of the house, and at night they would wage raucous territorial wars under our bedroom window.

Then, in mid-summer 1995, they vanished, never to be seen again, a fate that befell all the weka in our region. Capable of killing rats and even stoats, weka were thought to be able to hold their own against the barrage of human challenges thrown into their rambunctious path. Yet their mysterious disappearance continues to spread across the landscape. Having named our New Zealand home 'Weka Wilds', we have not yet finished mourning their absence in the now-silent nights which not so long ago rang with their piercing 'come-and-get-me-if-you-can' cries.

RIGHT The crystal-clear Cobb River emerges from pristine beech forest in Kahurangi National Park.

ABOVE Lake Brunner, fringed in flax and wild forests, is home to many dabbling ducks.

RIGHT Now surviving only on subantarctic islands, the New Zealand snipe was once widespread in wet places.

ABOVE Wetlands permeate the forests along the West Coast. Here, water cascades down the a steep undercut creek bank.

LEFT The banded dotterel frequents river flats as well as soggy subantarctic herbfields.

RIGHT Surrounded by rain-drenched forests, tranquil Lake Moeraki is favoured by waterfowl.

BELOW LEFT AND RIGHT Unlike introduced mallards, the native grey duck shows no colour difference between sexes; elegant stripes on its face and the iridescent green speculum are distinguishing features.

BELOW MIDDLE The endemic New Zealand scaup is an adept diver, preferring the deep, cool lakes of the high country.

RIGHT The Waimakariri River flowing out of Arthur's Pass National Park is a classic braided river, where thin ribbons of water meander sinuously across wide shingle plains, carrying rock debris away from the youthful mountain ranges into the plains.
BELOW Closely monitored and protected, an endangered black stilt, one of the rarest waders in the world, seeks cadis fly larvae in the braided outflow of the glacially fed Tasman River.

LEFT Black-billed gulls nest on an island on a small lake in the hill country of Central Otago.

ABOVE Rain water pours off the sheer walls of Milford Sound, forming a layer of fresh water on top of the sea.

RIGHT Along the Haast River Gorge, limpid waters cascade down the hard schist canyons clad in wild beech forest.

LEFT With strange fleshy baffles on the edges of its bill, the blue duck is a specialist of mountain torrents and fast-flowing rivers, where it leads a sedentary life diving amid turbulent eddies. Its numbers are dwindling through predation and habitat loss.

LEFT Water flows everywhere during a heavy rainstorm in Milford Sound, where annual rainfall is well over 6 m (20 ft).

ABOVE Lake Hawea is one of several large lakes that lie in the beds of vanished ice age glaciers along the eastern foothills of the Southern Alps.

BELOW Some of the clearest water in the world re-emerges at Waikoropupu Springs after travelling underground through miles of honeycombed limestone. Welling up at a rate of 14,000 litres (3,700 gallons) a second, this is one of the world's largest freshwater springs.

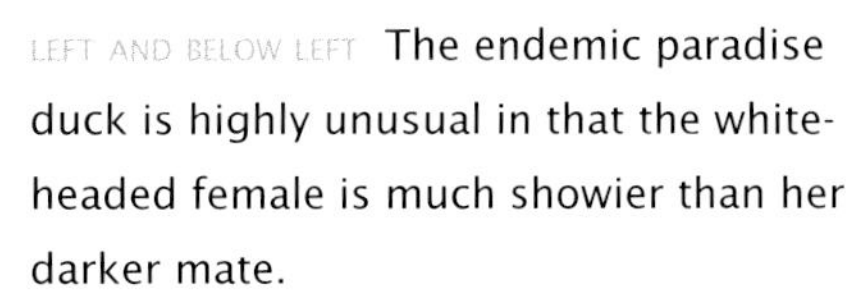

LEFT AND BELOW LEFT The endemic paradise duck is highly unusual in that the white-headed female is much showier than her darker mate.

RIGHT Feeding under a watchful motherly eye, a paradise duckling is boldly patterned.

BELOW Paradise duck pairs stay together for life, living in wet grassland habitats, such as here at Mount Cook National Park.

LEFT More graceful than the pampas grass introduced from South America, five species of toetoe grass grace swamps and coastal wetlands with their tall heads waving in the wind like flags.

RIGHT A pukeko sits on its nest in a beachside marsh. The birds breed in large family groups, protecting and raising the chicks communally.

BELOW A pair of weka pull worms from the muddy edge of a slow-flowing stream.

ABOVE The brown teal is a small endemic dabbling duck living in dense flax swamps and adjacent thickets. Territorial and highly aggressive, it is capable of killing rivals and larger duck species with ease. Yet it is no match for introduced predators, having dwindled in numbers to become one of the rarest ducks in the world. It is also unique in that pairs look after their young until they are fully grown.

RIGHT In contrast with its endemic cousin, the grey teal is nomadic, easily travelling across the sea from Australia.

LEFT A classic flax swamp.

FOLLOWING PAGES Taking advantage of low tide, weka venture out onto a Golden Bay beach to feed on marine invertebrates. They make dedicated parents, scouring the swamp for worms, insects and even nestlings to feed to their young, but forcibly evict their chicks from the territory when they grow up. Flightless and apparently vulnerable to disease or predation by wild ferrets, weka have mysteriously disappeared from the Golden Bay region in the last decade.

OPPOSITE (CLOCKWISE FROM BOTTOM LEFT) Wetland birds: blue duck, Australasian coot, grey teal, brown teal, New Zealand scaup, pied stilt, banded dotterel, spurwing plover, pukeko, variable oystercatcher, pied oystercatcher, paradise duck, New Zealand shoveler, grey duck, weka and white heron (ranging worldwide as the great egret).

ABOVE Shellfish-rich tidal flats in Abel Tasman National Park are a shorebird haven.

ABOVE RIGHT Pied stilts are common from seashore to inland pools.

RIGHT AND FAR RIGHT Weka scavenge along the shore on a penguin carcass and among marine life stranded by the tide.

7

OF WIND AND WAVE

THE WILD COASTLINES

Saving a whale must rank as one of my oddest experiences. Just after Christmas one year, the long days of the southern summer solstice lingering brightly, a pod of long-finned pilot whales seemed to become entrapped within the wide sandy sweep of Farewell Spit, which encloses Golden Bay. About 30 of these beautiful, enigmatic cetaceans — not true whales but actually oversize dolphins 6 m (20 ft) in length — were milling about in confusion, evidently trying to head west in spite of the shoaling sandspit. As they came closer into the shallows they converged into a tight cluster, a couple of very large bulls with a number of more slender cows and several small calves. Appearing to be conferring, they drew themselves together, squealing and whistling loudly for several minutes while rolling about and lolling vertically in the water with their strange bulbous heads emerging above the surface, their small pouting snouts and built-in fixed smiles showing. Then, their minds apparently made up, the entire group headed resolutely as one toward shore.

ABOVE Dusky dolphins pirouette in midair along the Kaikoura coast.

LEFT Sea spray rises from the wave-washed West Coast at Wharariki Beach, where hardy flax preside over storm-carved islets.

Along with several other small boats, we tried to race before them, zigzagging and circling to churn and froth the water, hoping the bubbles would head them off before they became stranded by the dropping tide. Time and again they veered out and away, but soon drew a semi-circle to repeat the entire performance a mile or so down the coast. It appeared they were quite aware that the wide open ocean from whence they had come lay not far to the west, but with the silty, featureless bottom rising gently toward the beach, their highly tuned echolocation abilities were foiled. With each attempt, the confused animals aimed doggedly in the direction of the Tasman Sea without swinging wide to clear the 35 km (20 miles) of wind-drifted sands that make up the treacherous spit. Believed to also make use of the earth's magnetic field as a navigation aid, a known local magnetic anomaly may have further disoriented them.

Some time after noon word came that the inevitable had happened. Caught by the receding tide, which leaves large areas of mudflat exposed, the pod had become stranded high and dry in the blazing sun. News travelled quickly, and people from far and wide arrived to try and do what they could to save the whales. First wet sheets were spread over the animals to keep their black backs from becoming sunburnt. A sprinkler system with small pump and long hose was set up to bring seawater from a tidepool to keep them cool. Then the long wait for the returning tide began.

Pinned to the ground by gravity and robbed of their constant, graceful mobility for the first time in their lives, the whales made for a sorry sight indeed. Pod members whistled and squeaked pathetically, unable to communicate with one another through air. Small calves lay alongside their lactating mothers whose milk was leaking uncontrollably from the pressure exerted by their own weight on the sand. Eyes shut against the blinding sun, they struggled to raise their heads instinctively for each breath. To ease their cruel discomfort, children busied themselves digging away the sand beneath the worst pressure points, or to free bent-over flippers. And when the afternoon flood finally did slowly return, each whale needed an attendant to hold it upright so it would not roll over in its struggle and, unable to right itself in the too shallow water, drown with its blowhole trapped below the surface.

By evening the weather had taken a nasty turn, a westerly gale now lashing cold rain into our faces. It was well past midnight before the tide was finally high enough to free the large animals from their hopeless predicament. The night was dark as coals and angry wavelets were breaking over my shoulders as I still hung on to 'my' whale — as did a couple dozen other volunteers — which had become so exhausted and seemingly resigned that it accepted my help peacefully. One arm under its chest and the other over its back, when I talked to it calmly it would stop struggling and allow me slowly to orient and push it toward the open sea. Another whale was thrashing feebly against my back. Finally, on a word of command we released all of our charges at once and quickly moved to stand in a line parallel to shore, effectively creating a human wall that we hoped the whales would use to orient themselves away from the coast.

Unique to New Zealand, the beautiful Hector's dolphin is one of the world's smallest, reaching a length of only 1.4 m (4½ ft).

Exhausted and shaking from cold, I returned to land as the whales vanished into the liquid night, back to their mysterious world out in the deep ocean. That 'my' whale had seemed to vest its trust in my efforts to succour it had a profound calming effect on me in return. Sightless and up to my neck in freezing water, I had felt only peace in the intimate company of such a massive animal, though I knew it was capable of knocking me senseless with one flick of its fluke. Oddly perhaps, I would have been far more nervous had I found myself pressed between dairy cows of similar bulk in the mundane setting of a barn, than I was that night sharing the dark waters with whales.

The seas around New Zealand abound with cetaceans, some of them barely known to us at all. Several of the world's rarest and least known beaked whales have been recorded in these waters. The year after the pilot whale stranding, a lone female pygmy right whale, the rarest of all the baleen — or great — whales, also stranded in Golden Bay. Almost nothing is known about its life habits other than it lives in the cooler regions of the Southern Hemisphere.

Dolphins thrive around all coasts as well as offshore. Common, bottlenose and the highly exuberant dusky dolphins, as well as the occasional family group of orcas, are features of most days on the water. On the other hand, the petite and beautifully marked Hector's dolphins, or tupoupou, found only in New Zealand and frequenting mainly the shallow, turbid waters of estuaries and river outflows around the South Island, are dangerously impacted on by human activity. Many become entangled and drown in recreational gillnets set near shore. A tiny remnant population of North Island Hector's dolphins, which were found to be genetically distinct and have been renamed Maui dolphins, are down to not much over 100 individuals, resident along the stretch of coast between New Plymouth and Hokianga Harbour only.

New Zealand boasts an astounding 18,200 km (11,300 miles) of coastline, representing just about every habitat imaginable, from sand dunes to sea cliffs, sheltered coves to wind-scoured headlands, deeply indented fiords to starkly jutting peninsulas.

Along the western shores the major effect of the West Wind Drift and the raw power of the Roaring Forties — some of the stormiest latitudes in the world — are felt to full effect. The coastlines are as wild and rugged as anywhere on the planet, gnarly sea stacks and wave-carved cliffs shouldering the full force of the turbulent ocean. In the far south, time-tested buttresses rise staunchly into swirling, rain-heavy clouds. Here riotous primordial forests tumble vertiginously into the sea, where ribbons of bull kelp swirl sensuously in perpetual motion. Glacier-carved fiords slice enormous gashes up to 40 km (25 miles) deep into the breathtaking mountains of Fiordland National Park, their sheer undersea walls harbouring unique, deep-sea marine life sheltered in the dim light beneath layers of tannin-tinted freshwater runoff.

An Antipodean wandering albatross scavenges on a dead cookie-cutter shark near the Kaikoura Peninsula.

The eastern shores are generally more sheltered and shelving, although headlands such as the Catlins Coast and Kaikoura Peninsula in the south, and East Cape and Cape Kidnappers in the north, exhibit their own brands of wildness. Likewise, the ancient volcanoes of Otago and Banks Peninsulas bear the marks of past glaciations through deeply breached calderas and jagged summit ridges. In the farthest northern reaches of the country run the seemingly endless sand dunes of Ninety Mile Beach. In the south, one of the strangest formations of sandy shores are the enigmatic Moeraki boulders of the Otago coast. Ancient mineral aggregates unearthed by wave action from the silty alluvial plain, these perfect spheres lay strewn along the beachfront as if huge cannonballs marked the vestiges of some long-forgotten giants' battle.

With each of these distinctive coastal habitats come different native species. Dunes and sandbanks of the north are favoured by the endangered New Zealand dotterel, and the even rarer fairy tern, now down to a mere few dozen birds. The much larger and more common Caspian tern seeks rougher shingle beaches. Rocky headlands attract nesting red-billed and southern black-backed gulls and black-fronted terns. Sea stacks and rugged islets are the premises of both pelagic seabirds like petrels and shearwaters, as well as coastal feeders such as Australasian gannets and the eye-catching, elegant endemic spotted cormorant or shag. Large tidal flats, most notably those of Kaipara and the Firth of Thames in the north and Farewell Spit in the south, attract vast concentrations of migrant shorebirds, none more numerous than the bar-tailed godwit, who travels some 10,000 km (6,000 miles) all the way from Siberia each southern spring.

Deep offshore waters are rife with cold-loving life from giant squid to lanternfish and frostfish. Rare leatherback turtles and giant sunfish work the mid-waters in search of 3-m (10-ft) long jellyfish. Basking sharks — the second largest fish in the ocean after whale sharks — use their cavernous funnel mouth to filter plankton from the surface layers.

A large undersea canyon running out from the Kaikoura Peninsula is famous for its rich upwelling waters that attract a wide array of cetaceans, from oddities like the sleek, finless right-whale dolphin to a stable population of sperm whales. These behemoths — the largest of all toothed whales — are almost all males up to 18 m (60 ft) in length, who commute to breed in tropical waters where the smaller females reside. New Zealand fur seals, or kekeno, are also common around Kaikoura, having made a remarkable comeback from the brink of extinction during the late 19th century. Once retrenched to the rugged south, new breeding colonies have reclaimed many rocky shores and islets around the South Island, and some are even recolonising the North Island. Still on the increase, their numbers have reached around 100,000, yet this figure is believed to represent only about 5 percent of the pre-sealing population.

Totally at home in the open ocean (they only need to come ashore to breed), the seals travel hundreds of miles offshore where they may dive down to 270 m (880 ft) for up to ten minutes in pursuit of their favourite prey, lanternfish. However, when hoki concentrate in the depth of the Hokitika Canyon off the West Coast to spawn each winter, this provides a bonanza upon which the seals rely, especially mothers with pups ashore. But their impressive diving prowess only barely brings them within reach of the spawning fish schools in the dark depths, so they often follow the large commercial trawlers that also converge here to snatch fish as the huge trawl nets are raised. Consequently, many seals become trapped in the nets and drown which, in the case of lactating cows, causes orphaned pups to starve on shore. Severely overfished, the collapsing stocks of hoki do not bode well for the survival of the seals in the West Coast region.

It indeed seems odd that in a country as conscious of its natural heritage as is New Zealand, few people are willing to consider that fish are wildlife too, with many species in severe decline due to over optimistic quotas and the rather small number of no-take marine reserves, where inshore species may breed unimpeded to allow stocks to recover. According to reports publicised by the Royal Forest and Bird Society, the annual figures for New Zealand's commercial fisheries are indeed staggering: 600,000 metric tons of seafood caught, using 120,000 bottom trawls, 90,000 dredge hauls, 20,000 openwater tows, plus a mind-boggling 54 million baited hooks set on long lines each year. Not surprisingly, many non-target animals also fall prey to this massive scale of human predation, including corals and other invertebrates on the seafloor, dolphins and sharks in mid-water and large numbers of petrels and albatrosses on the surface, all coldly and succinctly labelled 'bycatch'.

A yearling fur seal yawns after a siesta in the cool shade of shoreline boulders.

New Zealand waters represent some of the richest and most diverse seabird assemblages in the world, many found nowhere else. A record-breaking 14 of the world's 24 albatross species breed on New Zealand's offshore islands — including the subantarctic — far more than any other single country. Numerous petrels also nest throughout New Zealand, some deep in the lush forests where no seabirds would ever be imagined to venture, others restricted to tiny locations offshore. The large Westland petrel chooses the lush Paparoa Ranges for its nesting colonies, while Hutton's shearwater flies high into the Kaikoura Ranges to reclaim its nesting grounds amid snowdrifts each spring. The taiko, or magenta petrel, of the Chatham Islands, is acutely endangered with under 200 surviving only with the help of intense conservation efforts.

Just as New Zealand represents the centre of albatross diversity, penguins, too, are deeply linked with these ancient shores. The little blue penguin, korora, is the smallest of all penguins standing just 40 cm (16 in) high. Widespread around all main coasts, it feeds in inshore waters and comes ashore in small discreet flocks under the cover of darkness to nest in caves and hollows under boulders and logs. The stocky Fiordland crested penguin, or tawaki, on the other hand, confines itself to the wildest coasts of Westland, Fiordland and Stewart Island. Here they sneak ashore largely unseen and vanish into the deepest, steepest, wettest forest understory they can find. Favouring this wildest of environments, their range does not extend far around the bottom of New Zealand, where their place is taken on the east coast by a larger cousin, the yellow-eyed penguin, or hoiho. These unusually marked penguins select shoaling beaches and protected coves to venture ashore along the Catlins, Otago and Canterbury coasts, as well as Stewart Island where all three rub flippers in several places. The yellow-eyes do not form dense colonies as do many other species, but nest alone, secretively tucked away in dense coastal vegetation. Unfortunately, like so many other New Zealand creatures, they have suffered badly from predation by introduced mammals, loss of coastal habitat, and also fire. Recently, in some areas a mysterious disease claimed up to 30 percent of the 2004/2005 chicks. Sadly, all three species are listed as threatened or endangered.

Three more penguin species — the rockhopper, erect-crested and Snares crested — breed in the subantarctic islands south and east of the main shores. Of this total of six, four occur nowhere outside New Zealand. What's more, the fossil record shows 14 additional extinct species left traces of their past existence on these shores, including one that stood nearly as tall as a person.

The mysteries of New Zealand's strange and wonderful life forms, whether giants or dwarfs, create a vibrant web that endures, full of surprises and discoveries yet to be made. And just as we occasionally despair in the face of seemingly insurmountable challenges to ensure a safe future for this priceless natural treasure, fate appears to smile in delivering good news.

On January 25, 2003 a tiny black and white seabird was sighted in the Hauraki Gulf, the first live observation of the New Zealand storm petrel, missing in action for 150 years. Known only from fossil bones and three museum specimens collected in the 1800s, this encounter (and subsequent confirmed sightings) was akin to seeing the living dead, a bird whose nesting grounds have never yet been discovered to this day.

Similar happy news is starting to come from marine mammals. Not only are fur seals bouncing back from the brink of extinction in their tens of thousands, so too we may celebrate the return to New Zealand waters of some of the most threatened whales. Once ruthlessly exploited, critically endangered southern right whales — so named because they were the 'right' whale to hunt — are once again safely breeding in the recently gazetted Auckland Islands Marine Reserve, individuals venturing up the east coast of the South Island with increasing frequency as far north as Cook Strait.

Likewise, the largest animal ever to grace the face of the earth, the southern blue whale, plies western waters along the continental shelf. Savagely hunted until the 1960s, the Southern Hemisphere race is considerably larger than its already substantial northern relatives. These giants of giants can reach 30 m (100 ft) and 160 tons, engulfing around eight tons of krill per day, which they filter through their enormous sieve-like baleens. Their heart alone is about the size of a small car, and their arteries similar to storm drainpipes. At birth, a calf is over 7 m (20 ft) long and, fed on milk as rich as double cream, grows at a rate of about 90 kg (200 lb) per *day*.

For me, spotting several of these behemoths at close range, off Kaipara Head in the North Island and Kahurangi Shoals in the South Island, ranks as a heart-stopping moment. On a calm day their cavernous exhalation when breaking the surface to breathe can be heard nearly a mile distant, the mist of their warm breath towering 9 m (30 ft) high. Their mousy blue-grey, finely mottled backs glow pale turquoise like a shallow reef just under the surface, and when they plunge again their splendid, streamlined flukes some 6 m (20 ft) across rise in slow motion high above the waves to provide downward momentum, water cascading off the trailing edge like a waterfall. I can only describe it as a quasi-religious experience. Fittingly, blue whales produce the loudest sound of any animal on earth, a long subsonic moan rising to an incredible 180 decibels. Without speeding them up, the frequencies are too low for our ears to register and are believed to travel extraordinarily well across the deep ocean, enabling the whales to communicate over vast distances. Some theories even posit that they may keep contact across entire ocean basins or perhaps even from one ocean to another. When it comes to New Zealand's natural world and its many strange and mysterious life forms — whether on land or at sea — the age of discovery has only just begun.

ABOVE A shy yellow-eyed penguin walks across the sward to its nest in dense scrub.

BELOW Once hunted nearly to extinction, blue whales frequent the western offshore waters.

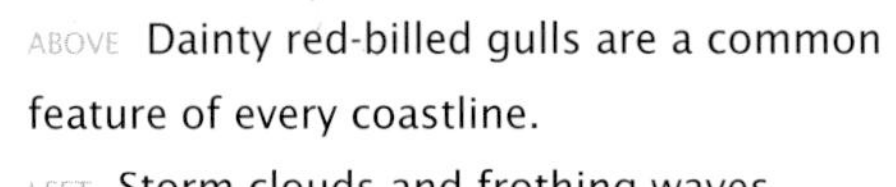

ABOVE Dainty red-billed gulls are a common feature of every coastline.

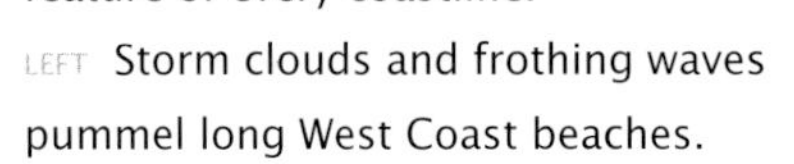

LEFT Storm clouds and frothing waves pummel long West Coast beaches.

ABOVE In the Hauraki Gulf, the tiniest of islets sports its own tree.

BELOW Driven by fierce westerly winds, regular sets of waves from the Tasman Sea roll in along the Paparoa coast.

ABOVE A pair of elegant white-fronted terns court atop a granite boulder; the male presenting his prospective mate with a fish offering.

BELOW Common dolphins break the surface of a velvet sea shortly after sunrise.

RIGHT A sandstone sea stack makes a formidable citadel for an offshoot of the large gannet colony at Muriwai Beach on the North Island's west coast.

ABOVE LEFT A double rainbow over Golden Bay, South Island.

LEFT AND ABOVE RIGHT Larger than other species, bottlenose dolphins cruise through calm coastal waters bordering Abel Tasman National Park.

ABOVE LEFT AND RIGHT The spotted shag is an elegant small cormorant found only in New Zealand. Building nests of seaweed on cliff ledges, pairs acquire fine crest feathers and stunning facial colours during courtship.

RIGHT Small islands along the Abel Tasman National Park shoreline are used by seabirds for both roosting and nesting. BELOW An adept diver, a spotted shag struggles to gain lift over a windless sea.

ABOVE Heavy swells begin to break as they reach the shallows.

LEFT AND BELOW The pancake limestone rocks of Punakaiki are deeply carved by wave action, creating magnificent blowholes when the tide is high.

BELOW With their oversize webbed feet, spotted cormorants are better swimmers than fliers but must come ashore to roost as their feathers have poor waterproofing.

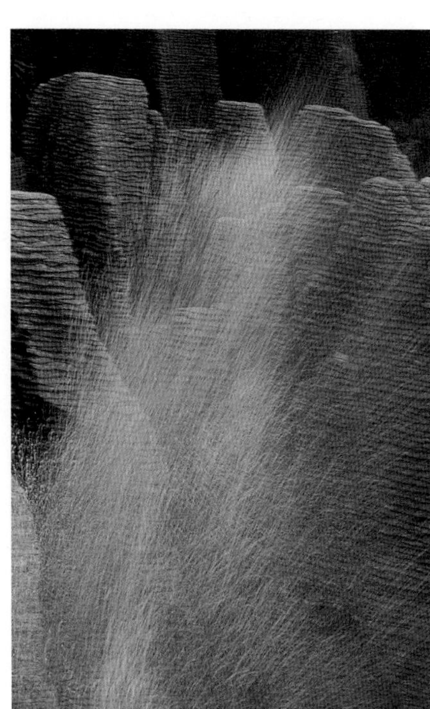

ABOVE AND LEFT Cruising over the ocean on strong wingbeats, gannets are splendid plunge-divers; dropping from 30 m (100 ft) or more at speeds up to 145 kph (90 mph), the bird hits the water like an arrow, diving down at least 8 m (26 ft) in pursuit of sardines and other fish.

RIGHT Australasian gannets nest on many sea stacks and rocky islets, especially around the North Island, but only a few colonies are found on the main shore. Here birds nest on bluffs near Muriwai Beach, west of Auckland.

BELOW Bull sperm whales come to feed on squid in the deep waters of Kaikoura's undersea canyon, whereas females, pictured here, tend to remain in more tropical seas.

Steep underwater canyons near the shore bring rich marine life close to the spectacular Kaikoura coast.

ABOVE In late winter large flocks of Hutton's shearwaters gather together in rafts before heading inland into the snowy Seaward Kaikoura Ranges to nest.

RIGHT Fur seals haul out on the rocky tip of the rugged Kaikoura Peninsula, with bulls much larger than cows.

Many dolphin species are part of the retinue of marine mammals frequenting Kaikoura's waters.

ABOVE A pair of common dolphins swims in synchrony as they mate just below the surface.

RIGHT One of the world's rarest, the endangered Hector's dolphin numbers only about 3000–4000, living close to shore in pairs or small pods.

BELOW Acrobatic swimmers, reaching speeds of 35 kph (22 mph), large groups of dusky dolphins visit coastal waters, although they also travel well offshore.

ABOVE *Olearia* daisies grow just above the tideline in Fiordland.
BELOW Delicate invertebrates flourish beneath a surface layer of fresh water resulting from heavy rain runoff in Fiordland's deep sounds.

ABOVE AND LEFT The rare Fiordland crested penguin comes ashore in winter to nest deep in the forest along the rugged southern West Coast. Crested penguins lay two eggs of uneven size, with only the chick from the largest egg surviving. As they grow, the youngsters flock together while their parents commute to sea to feed.
MIDDLE LEFT The little blue penguin does not stray far from shore, normally making short dives to 10 or 20 m (30–65 ft) but sometimes descending to 60 m (195 ft). It comes ashore at night to nest in dense vegetation.

RIGHT The coastlines at the top of the South Island enjoy varied geology and exceptionally calm weather. Golden sands and granite pillars are hallmarks of the Abel Tasman National Park.
BELOW LEFT Just off Farewell Spit, a mother and calf long-finned pilot whale break the surface together to breathe.
BELOW RIGHT Sunset over Wharariki Beach, at the top of the South Island's west coast.

ABOVE The Moeraki boulders, about 2 m (6½ ft) in diameter, 2 ton in weight and strewn like giant cannonballs on an Otago beach, are 60 million-year-old aggregations of lime-rich minerals that formed on the sea floor.

OPPOSITE Some 18,000 km (11,300 miles) of coastline, plus rich seabeds and variable ocean currents offshore, offer a vast diversity of habitats for an eclectic list of marine wildlife. (CENTRE) Australasian gannet; (CLOCKWISE FROM TOP LEFT) Buller's albatross, Campbell albatross, Fiordland crested penguin, little blue penguin, Antipodean albatross, yellow-eyed penguin, common dolphin, Hector's dolphin, orca, dusky dolphin, New Zealand fur seal, grey-faced petrel, spotted shag or cormorant.

ABOVE A yellow-eyed penguin feeds its hungry chick hidden from view beneath coastal scrub.

LEFT Denizens of the southeast and subantarctic, timid yellow-eyed penguins returning from the sea at the end of the day pause as they crest a ridge heading for their separate nests concealed some distance inland.

Appendix: List of Species

Maori names are given in paranthesis unless they are in primary use.

adzebill *Aptornis* spp.
alpine buttercup *Ranunculus* spp.
Antipodean wandering albatross *Diomedea antipodensis*
Archey's frog *Leiopelma archeyi*
Australasian bittern *Botaurus poiciloptilus*
Australasian coot *Fulica atra australis*
Australasian gannet (takapu) *Morus serrator*
Australian brush-tailed possums *Trichosurus vulpecula*

banded dotterel (pohowera, tuturiwhato) *Charadrius bicinctus bicinctus*
banded rail *Rallus philippensis*
bar-tailed godwit *Limosa laponica*
basking shark *Ctorhinus maximus*
beech *Nothofagus* spp.
bellbird (korimako) *Anthornis melanura*
bidibidi *Acaena* spp.
black-billed gull *Larus dominicanus*
blackbird *Turdus merula*
black daisy *Leptinella atrata*
black-eyed gecko *Hoplodactylus kahutarae*
black-fronted dotterel *Elseyornis melanops*
black-fronted tern *Sterna albostriata*
black petrel (taiko) *Procellaria parkinsoni*
black stilt (kaki) *Himantopus novaezelandiae*
blue duck (whio) *Hymenolaimus malacorhynchos*
bottlenose dolphin *Tursiops truncatus*
brown creeper (pipipi) *Mohoua novaeseelandiae*
brown teal (pateke) *Anas chloritis*
bush wren *Xenicus longipes*

cabbage tree (ti kouka) *Cordyline* spp.
Caspian tern (taranui) *Sterna caspia*
cave spider *Spelungula cavernicola*
cave weta *Pharmacus* spp.
celery pine *Phyllocladus alpinus*
centipede (weri) *Cormochephalus rubriceps*
common blue butterfly *Zizina otis labrados*
common dolphin *Delphinus delphis*
common tree weta *Hemideina thoracica*
cookie-cutter shark *Isistius brasiliensis*
coprosma *Coprosma* spp.
crown fern *Blechnum*

dabchick (weweia) *Poliocephalus rufopectus*
dusky dolphin *Lagenorhynchus obscurus*

erect-crested penguin *Eudyptes sclareti*
European stoat *Mustela erminea*
Eyles's harrier *Circus eylesi*

fairy tern *Sterna nereis*
fantail (piwakawaka) *Rhipidura placabilis*
fernbird (matata) *Bowdleria punctata*
Fiordland crested penguin *Eudyptes pachyrhnchus*
five-finger *Pseudopanax* spp.
flax (harakeke) *Phormium* spp.
flax snail *Placostylus ambagiosus*
flightless teal *Anas aucklandica nesiotis*
forest gecko *Hoplodactylus granulatus*
frostfish *Lepidopus caudatus*

giant crested grebe *Podiceps cristatus*
giant squid *Architeuthis longimanus*
great spotted kiwi (roa) *Apteryx haasti*
grey duck (parera) *Anas superciliosa*
grey teal *Anus gracilis*
grey warbler (riroriro) *Gerygone igata*

Haast eagle *Harpagornis moorei*
harlequin gecko *Hoplodactylus ralciurae*
Hector's dolphin (tupoupou) *Cephalorhynchus hectori*
Hochstetter frog (pepeke) *Leiopelma hochstetteri*
hoki *Macrcronus novaezelandiae*
huhu grub (longhorn beetle larvae) *Prionoplus reticularis*
huia *Heteralocha acutirostris*
Hutton's shearwater *Puffinus huttoni*

kahikatea *Dacrycarpus dacrydioides*
kaka *Nestor meridionalis*
kakapo *Strigops habroptilus*
kaki *Himantopus novaezelandiae*
kanuka *Kunzea ericoides*
kauri *Agathis australis*
kauri snail *Paryphanta* spp.
kea *Nestor notabilis*
kidney fern *Trichomanes* spp.
kiekie *Freycinetia baueriana* ssp. *banksii*
kiwi *Apteryx australis*
kokako *Callaeas cinerea*
korikori *Ranunculus insignis*
korimako (bellbird) *Anthornis melanura*
kowhai *Sophora tetraptera*

lancewood *Pseudopanax crassifolius*
land snails *Powelliphanta* spp.
lanternfish *Myctophum punctatum*
laughing owl *Sceloglaux albifacies*
leatherback turtle (honu hiwihiwi) *Dermochelys coriacea*
lesser knots *Calidris canutus canutus*
little blue penguin (korora) *Eudyptula minor*
little spotted kiwi *Apteryx owenii*
long-tailed bat *Chalinolobus tuberculatus*

magenta petrel (taiko) *Pterodroma magentae*
makomako *Aristotelia* spp.
mamaku *Cyathea medullaris*
manuka *Leptospermum scoparium*
Maori onion *Bulbinella* spp.
marsh crake *Porzana pusilla affinis*
masked lapwing *Vanellus miles novaehollandiae*
matagouri *Discaria toumatou*
Maud Island frog *Leiopelma hamiltoni*
Maui dolphin *Cephalorhynchus hectori*
moa, giant *Dinornis giganteus*
morepork (ruru) *Ninox novaeseelandiae*
mountain cabbage tree *Cordyline indivisa*
mountain daisy *Celmisia* spp.
Mt Cook lily *Ranunculus lyallii*
mudfish *Neochanna* spp.
myrsine *Rapanea guinensis*
native clematis *Clematis virginiana*
New Zealand dotterel *Charadrius aquilonius* (North Island); *C. obscurus* (South Island)
New Zealand forest falcon (karearea) *Falco novaeseelandiae*
New Zealand fur seal (kekeno) *Arctocephalus forsteri*
New Zealand pigeon (kereru or kukupa) *Hemiphaga novaeseelandiae*
New Zealand pipit (pihoihoi) *Anthus novaeseelandiae*
New Zealand scaup *Aythya novaeseelandiae*
New Zealand shoveller *Anas rhynchotis variegata*
New Zealand snipe (hakawai) *Coenocorypha aucklandica*
New Zealand storm petrel *Pelagrodoma marina*
nikau *Rhopalostylis sapida*
North Island brown kiwi *Apteryx mantelli*
North Island robin *Petroica australis longpipes*
North Island tomtit *Petroica macrocephala toitoi*
Northland green gecko *Naultinus grayii*
Norway rat *Rattus norvegicus*

Okarito brown kiwi (rowi) *Apteryx rowi*
orange-fronted parakeet *Cyanoramphus malherbi*
orca *Orcinus orca*
ourisia *Ourisia* spp.

paradise shelduck (putangitangi) *Tadorna variegata*
penwiper *Kalanchoe marmorata*
pied oystercatcher *Haematopodidae ostralegus*
pied stilt *Himantopus himantopus*
pilot whale *Globicephala* spp.
pohutukawa *Metrosideros excelsa*
poroporo *Solanum aviculare*
pukeko *Porphyrio porphyrio*
pygmy button daisy *Leptinella nana*
pygmy right whale *Caperea marginata*

rata *Metrosideros Robusta*
raupo *Typha orientalis*
red deer *Cervus elaphus*
red-billed gull (tarapunga) *Larus novaehollandiae*
red-crowned parakeet (kakariki) *Cyanoramphus novaezelandiae*
rifleman (titipounamu) *Acanthisitta chloris*
right-whale dolphin *Lissodelphus peronii*
rimu *Dacrydium cupressinum*
robin (toutouwai) *Petroica australis*
rock wren *Xenicus gilviventris*
rockhopper penguin *Eudyptes chrysocome*
royal spoonbill *Platalea regia*
ruddy turnstones *Arenaria interpres*

saddleback (tieke) *Philesturnus carunculatus*
scree skink *Oligosoma waimatense*
shearwater (titi) *Puffinus*
shining cuckoo (pipiwharauroa) *Chrysococcyx lucidus*
short-tailed bat *Mystacina tuberculata*
sika deer *Cervus nippon*
silver fern (ponga) *Alsophila tricolor*
silvereye (tauhau) *Zosterops lateralis*
skink (mokopapa) *Oligosoma nigriplantare*
Snares crested penguin *Eudyptes robustus*
song thrush *Turdus philomelos*
South American pampas *Cortaderia selloana*
South Island kokako *Callaeas cinerea cinerea*
South Island pied oystercatcher *Haematopodidae ostralegus*
South Island robin *Petroica australis australis*
South Island saddleback *Philesturnus carunculatus carunculatus*
southern blue whale *Balaenoptera musculus*
southern rata *Metrosideros umbellata*
southern right whale *Balaena glacialis*
southern tokoeka kiwi *Apteryx australis*
spaniard *Aciphylla* spp.
sperm whale *Physeter macrorhynchus*
spotless crake *Parzana tabuensis plumbea*
spotted shag *Stictocarbo punctatus*
spurwing plover *Vanellus miles novaehollandiae*
Stead's bush wren subsp. of *Xenicus longipes*
stick insect, common green *Carauisus morosus*
stinkhorn *Phallales* spp.
stitchbird (hihi) *Notiomystis cincta*
stoat *Mustela erminea*
striped skink *Ctenotus robustus*
sunfish *Mola mola*
supplejack *Ripogonum scandens*
swallow (warou) *Hirundo* spp.

takahe *Porphyrio mantelli*
toetoe *Cortaderia conspicua*
tomtit *Petroica macrocephala*
torrent duck *Merganetta armata*
totara *Podocarpus totara*
tree fern (mamaku) *Cyathea* spp.
tuatara *Sphenodon punctatus*
tui *Prosthemadera novaeseelandiae*
tunnel web spider *Porrhothele antipodiana*
tusked weta *Motuweta isolata*

umbrella moss *Hypopterygium* spp.

vegetable sheep *Haastia pulvinaris*
velvet worm, peripatus *Peripatoides indigo*

weka *Gallirallus australis*
Westland petrel *Procellaria westlandica*
wheki tree fern *Dicksonia squarrosa*
white-faced heron *Ardea novaehollandiae*
white-fronted tern *Sterna striata*
whitehead (popokatea) *Mohoua albicilla*
white hebe *Hebe speciosa*
white heron (kotuku) *Egretta alba*
white-tailed deer *Odocoileus virgineanus*
wood rose *Dactylanthus taylorii*
wrybill *Anarhynchus frontalis*

yellow-eyed penguin (hoiho) *Megadyptes antipodes*
yellowhead (mohua) *Mohoua ochrocephala*

Index

Bold numbers reference captions & photographs.